HUMAN ANATOMY AND PHYSIOLOGY

EDUCATION IN ANATOMICAL SCIENCES

Human Anatomy and Physiology

Additional books in this series can be found on Nova's website under the Series tab.

Additional e-books in this series can be found on Nova's website under the e-book tab.

Education in a Competitive and Globalizing World

Additional books in this series can be found on Nova's website under the Series tab.

Additional e-books in this series can be found on Nova's website under the e-book tab.

HUMAN ANATOMY AND PHYSIOLOGY

EDUCATION IN ANATOMICAL SCIENCES

PAUL GANGULY
EDITOR

New York

For permission to use material from this book please contact us:
Telephone 631-231-7269; Fax 631-231-8175
Web Site: http://www.novapublishers.com

Additional color graphics may be available in the e-book version of this book.

Library of Congress Cataloging-in-Publication Data

ISBN: 978-1-62618-488-6

Published by Nova Science Publishers, Inc. † New York

This book is dedicated to Alfaisal University

CONTENTS

PREFACE

Education remains the major tool for community development. Investment in human resources through different forms of education, promote predictability and ensure higher performance. The book entitled "Education in Anatomical Sciences" is dedicated to Alfaisal University and demonstrates a state-of-the art learning tool by which knowledge and skills of anatomy can be transmitted from teachers to students. The readers therefore may find this preface as a useful link and resource to familiarize with the missions and visions of this new University developed through the King Faisal Foundation (KFF) over the years.

It gives a great honor for Alfaisal University and its college of medicine to endorse this book with its contents as it covers the latest tools in anatomic learning and written by leaders in medical education, a group most of which are members of Alfaisal University. Alfaisal University although a young university started in 2008 it comes as a product of KFF, the oldest foundation in the kingdom of Saudi Arabia and was started after the late king Faisal, a leader who worked hard to invest in education in Saudi Arabia and Middle East. The foundation inspired its vision from Alfaisal interest in education. He believed that human development has to be built on three elements, faith, knowledge and performance.

The King Faisal Foundation was established in 1976 by the sons of King Faisal of Saudi Arabia. Prince Mohammed was the Chairman of the Board of Trustees of the King Faisal Foundation. The Managing Director of the Foundation is Prince Khalid Al-Faisal. The foundation is one of the largest philanthropic foundations in the world and currently Prince Bandar Bin Saud is its chief executive officer. The foundation presents an annual prize, King Faisal International Prize (KFIP), to "dedicate men and women whose contributions make a positive difference". The foundation awards prizes in Service to Islam, Islamic studies, Arabic Language and literature, Science and Medicine. It comes next to Nobel prize both in grade and value. in fact several KFIP winners were awarded Nobel prize after they received awards from KFIP. The foundation also established King Faisal schools that provide American diploma and International baccalaurean degrees in general education. It has also the King Faisal Center for Research and Islamic Studies which was established in 1983. Based in Riyadh, the center is an independent, not-for-profit institution financed by the Foundation. The center studies issues affecting the Arab and Muslim worlds, and organizes a number of conferences each year. Its Library collection covers the fields of Islamic Studies and Islamic Civilization. At present, there are more than 250,000 volumes in Arabic and Latin languages. There is also a collection of rare and out-of-print periodicals and more than half million articles. A separate audio-visual research library has more than 11,000 audio and videocassettes, slides, and films, which contain lectures and documentary material related to the Center's various activities and interests. The foundation has an active scholarship program that support students in both undergraduate and post graduate education. Alfaisal University represents the latest achievements of the foundation. It is indeed a private non-profit institution which aspires to be a world-class research university committed to creation, dissemination and application of knowledge in the fields of business, engineering, life sciences and medicine, and to the development of knowledge-based economies. The college of medicine which is affiliated with the Partner Harvard Medical International (PHMI) has positioned itself to be one of the leading institutes in the region. Its high admission criteria for national and international students, its high caliber faculty and its state of art facilities are steps toward top international ranking.

Medical education is now a well-established science aiming to improve learning process in teaching medicine. It has developed dramatically over the last few decades for both basic science and clinical components of medical education. Professor Paul, the Editor of this book and the chair of the department of anatomy and cell biology at the college have managed to work with imminent faculty to produce this book. His interest in medical education and to promote teaching in Anatomy to meet the latest innovation and technology was the main drive to bring this high quality book. The book will contribute positively in promoting teaching anatomy not just at Alfaisal University but rather globally.

We would like to congratulate all authors and contributors of the book for their excellent work and dedication to enhance medical education and I am sure that the book will be a leading one in its field.

Khaled Al-Kattan, Dean

Wael Al-Kattan, Director, Academic Affairs

College of Medicine
Alfaisal University, Riyadh
Kingdom of Saudi Arabia

December 28, 2012

In: Education in Anatomical Sciences
Editor: Paul Ganguly
ISBN: 978-1-62618-488-6

Chapter 1

ANATOMICAL EDUCATION IN THE PAST: LESSONS FROM HISTORY

Paul Ganguly
College of Medicine, Alfaisal University, Kingdom of Saudi Arabia

ABSTRACT

Since anatomy is the language of medicine, it is perhaps important to browse through historical perspective so that the evolution of medicine is clear to many of us. The gross anatomy, microscopic anatomy, embryology and neuroscience represent the core knowledge of education in anatomical sciences and lessons from history can only improve our teaching learning outcome. The book, "Education in Anatomical Sciences" is intended to answer most of the challenges that we face at the present time. This is complicated by the fact that many of us are confused with the exact position of anatomy in a given curriculum as many medical schools have adopted an integrated approach. If we have to continue teaching an important subject such as anatomy to the medical students, we must be innovative in terms of our approach of teaching in the presence of decreased contact hours to fulfill the curricular need of more integration. It is perhaps true that 21st century medical educators now believe that anatomy is useful once it is clinically oriented and integrated with other subjects. Thus the present chapter not only highlights the historical perspectives but addressed areas that are expanded in the subsequent chapters of this book to improve the essence of teaching learning outcome in anatomy.

1. INTRODUCTION

Hippocrates, the great physician, would have liked to say "thank you" for all the efforts of the anatomists in the past because of the enormous achievements that are now implemented in anatomical sciences education. The discipline of anatomy has a pivotal influence on the history of medicine as it serves almost as the language of medicine. The biological science that has been embedded within this subject has not been changed till today but the modalities that we use to understand anatomy have been significantly changed. The entire human body can be now displayed in the digital mode and one can dissect the body without involving human cadaver. Anatomy is now well integrated in the developments in molecular medicine, pathology, neuroscience, embryology, genetics, lab technology and clinical sciences and thus the modern medicine is thriving.

2. Anatomy-Historical Perspective

Anatomy is an ancient medical science taught since antiquity. I cannot start the chapter without acknowledging many legends in the history who worked hard to pave the pathway to modern medicine. The birth of biology probably started around Greek philosophers and scientists such as Alcmaeon and Empedocles (5th century BC). Early in the third century BC, Herophilus and Erasistratus, surgeons in Alexandria make the first scientific studies to understand anatomy in humans convicted with crimes [1]. Aristotle emerged around the same time as one of the most well-known natural philosophers who provided many opinions of human body. The father of medicine however, is Hippocrates II, who first wrote about human anatomy. Hippocrates advocated also the importance of the relationship between physician and the patient [2-4]. Till today we value his views in medicine as Hippocratic oaths remind us ethical implications and honesty in this noblest profession. In early AD, pioneering work of Galen is well-known to us for many facets of human anatomy. Galen's dissection on animals produced several anatomical tracts that are still valid and important [1-4]. In about fifteenth century AD, Leonardo da Vinci had series of anatomical drawings. The drawings brought immediate attention for accuracy of the scientific facts related to bone, muscles, embryology and internal organs. In the middle of sixteenth century, Vesalius a young medical student from Brussels participated in many dissections to clarify/disprove Galen's hypothesis [5]. Arabic scholar Ibn al-Nafi's disputed aspects of Galen's views, providing a model that seems to imply a form of cardio-pulmonary circulation in his Commentary on Anatomy in Avicenna's Canon. Al-Nafis stated that "blood moved from the heart to the lungs, where it mixed with air, and then back to the heart, from which it spread to the rest of the body" [6]. The greatest breakthrough came when William Harvey from England in seventeenth century indicated the role of heart and the circulation of blood. At the same time, Malpighie discovered microscopes that revolutionized the concept of Biology and the scientists were able to find smaller structures including capillaries. Dutch researcher, Anton van Leeuwenhoek subsequently modified microscopes for general use. The modern science of cell structure started with careful work of Felix Dujardin, Robert Brown, Matthias Schlelden and Theodor Swann in the nineteenth century. For more information on the history of medicine, the readers are advised to look at other excellent articles [7-13]. Many of the subsequent chapters have also historical perspectives that are well described.

The history of medicine and anatomy is incomplete if one does not also acknowledge the contribution from several countries that played major role in discovering and identifying various human structures involved in the disease process. For example ancient Egypt developed early part of medical education well described by Edwin Smith Papyrus. The detail of examination, diagnosis, treatment and prognosis of various ailments has been known to us through these ancient scripts. Babylonian medicine similarly expresses the concept of professional skill that also echoed in ancient India and China. Greek and Roman medicine put forward a concept of community medicine that is still the choice in a medical curriculum. Religious books have often strengthened these facts.

3. Anatomical Education in the Past: Gross Anatomy, Embryology, Histology and Neuroscience

Over the centuries, knowledge of anatomy remains a great influence on the basic principles of surgery. We all know that the word "anatomy" is derived from the Greek word "anatom" meaning to cut up or to cut repeatedly. Once again, Greeks showed rigorous efforts to understand the working of the living body that can translate into a structured system. Till today, surgical maneuver without understanding anatomy of the body is a futile exercise and is not recommended in any curriculum development. In the early 1900s, one fifth of the medical curriculum in the United States was for gross anatomy course, with more than 800 hour of lectures and laboratory-based teaching time [14]. Between the years 1930 and 1980, anatomy curricula did not change much in the entire world and remained static, as the teaching methodologies were mostly based on traditional technology. Anatomy education during this period was also conducted without much clinical context. However, in the 1990s, medical educators felt that the anatomy course could be more interesting and there was a need to move this subject towards clinical integration. In fact, such a move started early (1969) in isolated way in places like McMaster University, Canada where anatomy was integrated in a problem-based curriculum (PBL) [15]. The impact of innovative curriculum such as PBL was not felt until in recent years where there have been several changes in the way basic science and, in particular, how anatomy curricula have been structured [14]. There has been a shift in emphasis in increasing integration of the basic and clinical sciences, towards decreasing contact time, increasing the participation of clinical faculty in basic science teaching, and encouraging the use of technological and electronic advances [16]. This allowed to an increased use of more interactive teaching and learning strategies, including team-based [17], case-based, and problem-based learning [15] rather than traditional didactic lecture-based learning.

The classical paper published by Drake et al. [14] indicated that trend in gross anatomy teaching is decreased significantly over the recent years in USA. This also reflected in laboratory time that the students spend in dissection/prosection labs. In contrary, microscopic anatomy courses showed minimal changes and in reality there was a significant increase in use of virtual microscopy in microscopic anatomic course. Both neuroscience and embryology labs have been significantly reduced to cater towards the need in modern anatomy curriculum.

4. Anatomy in the Context of Medicine: Lessons from History

If we have to continue teaching an important subject such as anatomy to the medical students, we must be innovative in terms of our approach of teaching in the presence of decreased contact hours to fulfill the curricular need of more integration. I remember the fact when I was a medical student long time back. We were almost forced to learn all the structures in detail as the anatomy course was spanned across two years of the curriculum. The assessment system did not integrate much of clinical part as recall knowledge was the main emphasize of the course. We as students unnecessarily spent time to understand the difference between the left and right patella. Obviously we learnt the lesson from the past that a sound teaching of anatomy demands a good curriculum map. The curriculum outcome which is dependent on knowledge, skill and attitude is also directly related to this map and anatomy only plays a role in integration of the subjects. Thus in designing the 21st century anatomy curriculum and resource center one can take care of issues primarily geared towards advancement in medical education that is primarily clinically oriented. The anatomy resource center plays a major role and can be designed to allow the students to have self-or team-based learning. Accordingly, the center may develop modules such as musculoskeletal, cardiac, respiratory, gastrointestinal, reproductive, renal, nervous and head and neck. Each module can be arranged according to themes of a system (within a carrel). For example, when the students are exposed to gastrointestinal disease, one of the important themes that the students are likely to encounter is "absorption of ingested food materials". Accordingly, the students identify the need to study the structure function correlation of the intestine. The carrel in the module can be constructed to address this issue and will contain a) models of intestine b) wet specimens of various parts of intestine c) radiographs to match these parts d) labeled plastinated specimens e) CT/MRI to highlight the different positions of the intestine and their relationship with other organs f) histology slides to identify features of intestine g) diagrams to highlight peritoneum and the blood vessels and their arrangement h) simple drawings to emphasize embryology (mid gut rotation) and its anomalies and how they can affect the normal position of the intestine i) small write up of the physiology of intestine j) a diagram highlighting malabsorption because of dysfunction of liver, gall bladder, pancreas k) various charts related to biochemistry l) pathology specimens of intestine m) computer assisted program and n) some questions and answers for self-directed assessment and study. A structured program of living anatomy including ultrasound in the clinical professional skills curriculum can strengthen such an endeavor.

Did we learn anything from the mistakes in the past? If the answer is yes then there are many improvements accomplished over the teaching techniques of this discipline: 1) it cannot be taught in an isolated way 2) it has to be integrated to other subjects 3) the lab should be organized to allow such integration 4) it should be more clinically oriented 5) the dissection time spent by each student should be carefully balanced to see if they can get the same benefits by using other modalities such as digital techniques 6) the assessment system should accurately reflect integrated clinical anatomy and finally, 7) continued medical education workshop as a part of faculty development is integral and key to success of these changes. The question arises as to what extent can we do these changes? The new medical schools have certainly the advantages as they can start with 21st century concepts. The old medical schools have to be aggressive as the curricular changes are not easy to accomplish. Faculty will have resistance in terms of cut back of resource timing as well against demand for innovative teaching techniques that are not didactic but more of interactive sessions. Once again, faculty development should dictate the need of these changes in medical education.

5. Challenges that We Face

The anatomy as a subject grew up around the medical school. For a century it trained physicians and graduate students who have innovative ideas in medical education and transformed anatomy into a modern science that caters towards our need in understanding what we termed as "clinically oriented anatomy". The time has come when I felt that all of us needed a book to address the challenges that we face. We seek help from the leading medical educationists in anatomy to formulate a state-of-the-art anatomy book dealing with medical education. The basic question such as "how do we best teach anatomy" remains a pivotal influence in structuring this book. The answer to this question obviously lies in a) curriculum development b) method of teaching such as traditional teaching (lecture based) versus team-based or problem-based learning c) organization of the anatomy resource center c) how active are we in understanding modern medical education e) 21st century modalities that we use to teach medical students f) assessment system and formulation of questions that accurately reflects the curriculum and g) strategies to accommodate students' perception of teaching-learning outcome of anatomy.

Certainly the proudest part of the history of anatomy and medicine is not recorded on memorial stones, prestigious lectures or in famed professionalism. It lies in the service of physicians, medical professors and support staff engaged in rigorous activities around the world working tirelessly to bring the subject of anatomy closer to the heart of medical or graduate students. The book, "Education in Anatomical Sciences" is intended to answer most of the challenges that we face at the present time. It should document the progress that we have made in this area and also paves a pathway to understand the issues in terms of where we are heading to. At the outset, however, the objectives of this project must be clearly understood. First, we have very few documents describing the aim and progress of anatomy in one place in the form of a book. Second, we need to address the very important point as "why do we need changes"? Third, we have to document the key issues in developing anatomy curriculum if we accept changes. Fourth, in order to reiterate the fundamental question, we must continue to ask -how best can we teach? Fifth, anatomical education in gross anatomy, histology, embryology and neuroscience should be revisited to strengthen clinically oriented anatomy. Sixth, we need exact guidelines to formulate multiple choice questions (MCQ) or objective structured practical examination (OSPE)and finally seventh, the teachers must be aware of the recent trends as well various modalities available for teaching 21st century medical and graduate students. Thus, the book for the first time addresses education in anatomy and deserves a great value to medical schools engaged in problem-based learning or integrated systems curriculum.

Conclusion

The proposed book "Education in Anatomical Sciences" is unique as the essential contents are put together allowing one to browse through anatomical knowledge on a daily basis. It should also satisfy anyone who believes that medical faculties must follow a system which is educationally sound. Developing an excellent anatomy curriculum, assessment system, anatomy resource center are key to success that will allow us to address "how do we teach anatomy"? The book will target medical educationists and students who may find easier to develop concepts in gross anatomy, embryology, histology and neuroanatomy. Since planning for learning experiences, their implementation and student assessment are closely related activities, care is taken to develop a process for clinically-oriented multiple choice question in anatomy that satisfies the theme and objectives of anatomy. The issues related to laboratory activities have been also addressed so as to emphasize objective-structured practical examination that is integrated and clinically relevant during the early period of medical curriculum.

I personally thank all contributors who work seriously to complete the project. I sincerely hope that this book may help all medical educators working in the field of anatomy.

REFERENCES

[1] Adams, F. (1939). *The complete works of Hippocrates*. Adams, F. (Ed) Williams and Williams, Baltimore, 1-10.
[2] Von Staden, H. (1992). The discovery of the body. Human dissection and its cultural contexts in ancient Greece. *The Yale J Biol Med*, *65*, 223-41.
[3] Von Staden, H. (1989). *Herophilus: the art of medicine in Alexandria*. 1st ed. Cambridge University Press, Cambridge, MA, 1-666.
[4] Durant, W. (1939). *Aristotle, History of Animals*. In: Durant, W. (Ed.) The story of civilization. Simon and Schuster, New York. Pp410-671.
[5] Simeone, F. A. (1984). Andreas Vesalius: anatomist, surgeon, count palatine and pilgrim. *Am J Surgery*, 147, 432.
[6] West, J. B. (2008). Ibn al-Nafis, the pulmonary circulation, and the Islamic Golden Age. *J Appl Physiol.*, *105*, 1877-1880.
[7] Anson, B. J. (1908). *Moris human anatomy-a complete systematic treatise*. London, Mcgraw Hill, 1-20.
[8] Anson, B. J. (1908). *Moris human anatomy-a complete systematic treatise*. London, Mcgraw Hill, 1-20.
[9] Jackson, B. T. (1992). *An introduction to the of history of surgery*. In: The new Aird's companion in surgical studies. Churchill Livingstone, 1-26.
[10] Edelstein, L. (1967). The history of anatomy in antiquity. Temkin, O, Temkin, C. (Eds.) *Ancient medicine*. Johns Hopkins Press, Baltimore, 1-23.
[11] Malomo, A. O., Idowu, O. E. & Osuagwu, F. C. (2006). Lessons from history: human anatomy, from the origin to the Renaissance. *In J Morphol*, *24*, 99-104.
[12] Phillips, E. D. (1973). *Greek medicine*. Thames and Hudson, London, 4-33.
[13] Reveron, R. R. (2011). Marcello Malpighi (1628-1694), Founder of Microanatomy. *In J Morphol.*, *29*, 399-402.
[14] Drake, R. L., McBride, J. M., Lachman, N. & Pawlina, W. (2009). Medical education in the anatomical sciences: the winds of change continue to blow. *Anat Sci Edu*, *2*, 253-259.

[15] Chakravarty, M., Latif, N. A., Abu-Hijleh, M. F., Osman, M., Dharap, A. S. & Ganguly, P. K. (2005). Assessment of anatomy in a problem-based medical curriculum. *Clin Anat*, *18*, 131–136.

[16] Ganguly, P. K. (2010). Teaching and learning of anatomy in the 21st century: direction and the starategies. *The Open Med Edu J*, *3*, 5-10.

[17] Vasan N. S., DeFouw, D. O. & Compton, S. (2009). A survey of student perceptions of team-based learning in anatomy curriculum: Favorable views unrelated to grades. *Anat Sci Educ*, *2*, 150–155.

In: Education in Anatomical Sciences
Editor: Paul Ganguly
ISBN: 978-1-62618-488-6

Chapter 2

WHY DO WE NEED CHANGES? ANATOMY IN THE 21ST CENTURY

Jennifer M. McBride* and Richard L. Drake#
Cleveland Clinic Lerner College of Medicine
of Case Western Reserve University, Cleveland, Ohio, US

ABSTRACT

The field of medicine is as dynamic as the scientific research and technological advancements it is built upon, and yet, the way in which medical education is structured has remained relatively unchanged for nearly a century.

This chapter discusses the importance of implementing change in medical curricula to mirror the changing landscape and knowledge base in the biomedical sciences.

INTRODUCTION

For over a decade "curricular reform" has been the buzz words in medical education. Defining what this entails is somewhat complex, as recommendations can range from simply modifying aspects of course content to a complete course restructuring and integration with other courses. Even more daunting to define is the process in which to best implement these changes as each schools needs are subjective and predicated upon faculty number, student class size, building/room availability, technological constraints and budget limitations. All of these issues aside, major changes in medical education haven't occurred since 1910, after Abraham Flexner's report *Medical Education in the United States and Canada* was released [1]. Flexner's report recommended the following: 1) Medical schools should be affiliated

* Notes on contributors: Jennifer M. McBride, PhD, Assistant Professor in the Department of Surgery and Director of Histology at the Cleveland Clinic Lerner College of Medicine of Case Western Reserve University, Cleveland, Ohio.

Notes on contributors: Richard L. Drake, PhD, FAAA, Professor in the Department of Surgery and Director of Anatomy at the Cleveland Clinic Lerner College of Medicine of Case Western Reserve University, Cleveland, Ohio.

with universities; 2) Admission standards should be increased; 3) Sciences are fundamental to medicine and this premise should be reflected in training; 4) An academic model of education and training should be followed; 5) Medical schools should strive for higher quality faculty and better learning environments/facilities; and 6) Clinical training should be structured and have specific goals and objectives. In the mid 80's, the Association of American Medical Colleges (AAMC) released a report which had made 5 specific conclusions regarding the education of future physicians. One of the conclusions emphasized curricular reform by addressing student acquisition of learning skills. The recommendations listed under this conclusion were: 1) Evaluating the ability to learn independently; 2) Reducing scheduled time; 3) Reducing lecture hours; 4) Promoting independent learning and problem solving; 5) Using appropriate evaluation methods; and 6) Incorporating information sciences [2]. A similar suggestion was put forth in *Assessing Change in Medical Education – The Road to Implementation* (ACME-TRI) survey report in 1993 [3]. Within the Educational Strategies section of this report the recommendation to foster self-directed learning and lifelong learning skills is discussed along with possible approaches to its implementation. More recently, the Liaison Committee for Medical Education (LCME) has included in their accreditation guidelines that "A medical education program must include instructional opportunities for active learning and independent study to foster the skills necessary for lifelong learning" [4].

The cumulative summary of these studies and report recommendations on medical education necessitate change. As the breadth of knowledge in the biomedical sciences continues to grow with research discoveries and technological advancements there is no longer a question of why we need to change our curricular structures it is now a question of how we can change.

METAMORPHOSIS OF ANATOMY INSTRUCTION

To understand the premise for changes to the instruction of gross anatomy, we need to take a brief glimpse into the past. Due to the regulations set forth after the Flexner report there was a dip in the ratio of physicians to the population from 173 per 100,000 to 125 per 100,000 [5]. While this ratio increased modestly over the next several decades a major shift wasn't initiated until release of the report from the Surgeon General's Consultant Group on Medical Education in 1959 [6]. By 1975, this report predicted that the United States would experience a shortage of approximately 40,000 physicians. In response, the government (Kennedy and Johnson administrations) successfully lobbied to use federal taxpayer subsidies for the expansion of new, and growth of current medical schools. Interestingly, the American Medical Association (AMA) emphatically resisted these changes. Despite the AMAs resistance, an additional 38 medical schools opened bringing the total up to 126 between the years of 1965-1980 with more than 15,135 students graduating annually compared to 7,409 in prior years [5, 7]. So what did these changes mean for gross anatomy? For pre-existing schools, the rapid growth in class size severely challenged the existing facilities. While the laboratory section of many basic science courses were abandoned in the interest of space, the anatomy courses maintained their laboratory portion albeit, within a reduced timeframe. The other basic science laboratories, now no longer used for class activities, were swallowed up by research groups and research administrative staff, a move which increased the institutions research funding levels. The addition of full time scientists to the mix of teaching faculty with scientific interests created a shifting interest in the priorities of medical schools to re-define academic scholarship as acquisition of research funding and publication of knowledge.

Occurring at the same time, teaching began to take on a committee approach, wherein several faculty members were each responsible for a small segment of a large course as opposed to one faculty member charged with the whole course. According to one survey, the average ratio of students to full-time gross anatomists in American and Canadian medical schools was 29 to 1 [8]. In addition to the designated anatomy faculty members, faculty studying other areas of anatomy also participated in the education of medical students to provide an interdisciplinary course with details on clinical concepts, histological organization and/or systemic function. Moving forward, in the 90s, a survey of American and Canadian medical schools reported that the majority of responding programs indicated that their curriculum was largely traditional in that they still maintained a classroom and laboratory component [9]. The format of the classroom or non-laboratory component varied widely, from lectures to small group learning and even independent study. Additionally, the responding schools reported that they were testing alternate teaching methods such as problem-based learning (PBL), computer-based teaching and independent study. Like the classroom component, the laboratory component also varied widely between institutions with the following typically used in some sort of combination: student driven dissection, instructor or student led prosection, models and specimens. While the basic structure of gross anatomy instruction seemed to be maintained, several responding schools noted that their number of lectures was being reduced as was dissection time in the laboratory. More recently, in 2002, a survey of allopathic and osteopathic medical schools in the United States indicated that laboratory dissection was still a major portion of their courses [10]. A similar survey conducted in 2009 reported the same conclusions [11]. Like the previously mentioned studies, these reports also found that the laboratory consisted of a combination of activities.

TECHNOLOGY AND ANATOMY INSTRUCTION

Technology in teaching and learning has grown exponentially and continues to change. From Kodachrome slides and projectors to personal computers and PowerPoint presentations and now portable technology like smartphones and wireless media devices, the possibilities for technology in education are infinite. At present, it is practically unheard of for a lecture or seminar to occur without the use of some aspect of technology. As each device has emerged, the ways in which they can complement teaching, studying and evaluation have been developed. Often times, these pairings have facilitated instruction, helped students acquisition of knowledge and introduced other methods of their use such as testing, question analysis and distance learning. In addition to the positive contributions of technology there are also negative aspects such as poor class attendance as lectures may be posted online following the session, decreased faculty-student and student-student interactions, and the cost of implementing and maintaining the technology of interest. In anatomy there are scores of software applications, websites and textbook based resources. With the plethora of resources a major challenge for students and faculty alike is finding resources that are accurate and high yield in terms of acquisition of knowledge. In the classroom and laboratory many gross anatomists introduce the students to human anatomy with the use of various imaging modalities such as: computed tomography (CT), magnetic resonance imaging (MRI), X-ray and ultrasound. Utilizing imaging resources in gross anatomy instruction not only introduces the student to another way of viewing human anatomy but also presents it within a clinical context. Other ways in which students can view human anatomy and within a clinical context are with the introduction of surgical instruments. Several gross anatomists coordinate hands on laboratory activities for the students with various scoping procedures [12]. Whether it's web-based, imaging based, or surgically oriented, technology certainly has a place in teaching gross anatomy. Like any teaching tool the appropriateness of its use should be explored prior to implementation.

Clinical Anatomy in the Curriculum

Historically, incorporation of clinically relevant material into instruction of gross anatomy has occurred in some degree. As course hours were reduced and reports emerged of anatomy faculty becoming scarcer, a greater shift towards incorporation of clinically and functionally relevant material became a focus of gross anatomy courses [8]. Perhaps now more than ever, the clinically relevant application of this material has become a focus, not just in gross anatomy but in medical education as a whole. The most palpable reason for this is the demand on medical school curriculums to address the colossal amount of knowledge emerging from various fields in the biomedical sciences. Additional motivating factors include a call for more clinical experiences and recommendations from the AAMC and LCME to reduce lecture hours and promote independent study/learning. In an effort to expose students to more clinical applications of anatomy, textbooks began to emerge in the 80's with significant amounts of clinical correlations. Many of the brief clinical correlates in these texts were paired with pictures of structure organization and their relationship to skin projections referred to as "surface anatomy". Other views of the structures were included as well with the inclusion of X-Rays and ultrasounds. As more advanced imaging modalities became available, the figures in the textbooks were updated to include the structures as viewed with CT, MRI and positron emission tomography (PET). Beyond the textbook, laboratory sessions have also had an increase in clinically relevant material. Clinicians are often included in the instruction of these sessions and occasionally, so are residents and fellows [13]. In addition to varied levels of student completed dissections many programs include surgical procedures in select sessions to emphasize the importance of anatomical structures and their relationships to one another.

Conclusion – Why Do We Need Change

While it may not be recognizable, the way in which gross anatomy has been taught is constantly changing. As discussed, accreditation standards, curricular restrictions and staff availability will always play a role in these changes but the greatest reason for the fluid nature of this subject, to its credits, are the constant advances in the biomedical sciences. We need to change because the field changes.

References

[1] Flexner A. 1910. Medical Education in the United States and Canada: A Report to the Carnegie Foundation for the Advancement for Teaching. Carnegie Bulletin 4. New York, NY: The Carnegie Foundation for the Advancement of Teaching. Available from: URL: http://www.carnegiefoundation.org/sites/default/files/elibrary/Carnegie_Flexner_Report.pdf [accessed 27 December 2012].

[2] AAMC. Physicians for the twenty-first century. Report of the panel on the general professional education of the physician and college preparation for medicine. *J. Med. Educ.* 1984; 59:1-208.

[3] ACME-TRI Report. Educating medical students: Assessing change in medical education – The road to implementation. *Acad Med* 1993; 68:S3-S46.

[4] LCME. 2011. Functions and Structure of a Medical School. Standards for Accreditation of Medical Education Programs Leading to the M.D. Degree. Available from: URL: http://www.lcme.org/functions2011may.pdf [accessed 27 December 2012].

[5] Starr P. *The Social Transformation of American Medicine*. New York: Basic Books; 1982.

[6] Bane F. Physicians for a growing America. *Report of the surgeon General's Consultant Group on Medical Education. U.S. Public Health Service Publication No. 709.* Washington, D.C., October, 1959.

[7] Ludmerer KM. *Time to Heal: American Medical Eduction from the Turn of the Century to the Era of Managed Care*. New York: Oxford University Press; 1999.

[8] Blevins CE, Cahill DR. Gross anatomy: Current courses, training programs, and prospective needs. *J. Med. Educ.* 1973; 48:264–270.

[9] Collins TJ, Given RL, Hulsboch CE, Miller BT. Status of gross anatomy in the U.S. and Canada: Dilemma for the 21st Century. *Clin. Anat.* 1994; 7:275–296.

[10] Drake RL, Lowrie DJ, Prewitt CM. Survey of gross anatomy, microscopic anatomy, neuroscience, and embryology courses in medical school curricula in the United States. *Anat. Rec.* 2002; 269:118–122.

[11] Drake RL, McBride JM, Lachman N, Pawlina W. Medical education in the anatomical sciences: The winds of change continue to blow. *Anat. Sci. Educ.* 2009; 2:253-259.

[12] Reidenberg JS, Laitman JT. The new face of gross anatomy. *Anat. Rec.* 2002; 269:81-88.

[13] McBride JM, Drake RL. Rewarding the resident teacher. *Anat. Sci. Educ.* 2011; 4:227-230.

In: Education in Anatomical Sciences
Editor: Paul Ganguly

ISBN: 978-1-62618-488-6

Chapter 3

KEY ISSUES IN DEVELOPING ANATOMY CURRICULUM

Akef Obeidat*
College of Medicine, Al-Faisal University, Riyadh, Kingdom of Saudi Arabia

ABSTRACT

Anatomy has been the cornerstone of medical education for hundreds of years and it has traditionally been delivered at the beginning of medical education to provide a basis for clinical training and practice. Anatomy as a subject has suffered as a result of its failure to evolve and adapt quickly enough. Under old-style medical training, anatomy courses were very detailed and students were expected to learn those details with little understanding of relevance. This necessitates the need for anatomy to re-invent itself. It should evolve to address the requirements of any subject in a medical curriculum and to reflect the continuing nature of medical education.

In spite of delayed changes, it is true that some progress has been made. There has been a move from passive, didactic, highly detailed courses towards functionally and clinically relevant courses irrespective of the method of teaching. Recently, anatomical sciences have been vertically integrated in the medical curriculum and teaching of anatomy is witnessing a collaborative effort of basic science experts and clinicians by increasing the clinical relevance of what is taught.

In this chapter, I will discuss the guiding principles and key issues for the development of an educational program in anatomical sciences within the context of new trends in medical education.

INTRODUCTION

Anatomy is essential for physicians who perform an invasive procedure on a patient, carries out emergency procedures, examines radiological imaging, performs a physical examination of a patient, refers a patient to another doctor, or explains a procedure to a

* E-mail: aobeidat@alfaisal.edu. akefobeidat@hotmail.com.

patient. These tasks are common to all branches of medicine. Medical students have to be prepared for such tasks by changing the way we educate students in medical schools. Departments in the preclinical years of traditional medical courses provided learning environments, often with heavy workloads and a lack of clear specific objectives and goals. Anatomy has historically been a corner stone in medical education regardless of specialty. Until recently, dissection and didactic lectures were main instructional methods to teach detailed anatomy.

Anatomy can help in shaping the future of medical education. In coping with the recent advancements in medical education, anatomy has the capacity to lead the process of re-shaping basic medical sciences to keep up with the fast paced changes in curricular structure. Such a move also reflects changes in medical practice, health care delivery systems and the rapid advances in basic and clinical biomedical research that have led to enormous improvements in the prevention, diagnosis and therapy of disease. This might even be of help to certain undergraduate medical courses facing the threat of being externally dismantled, as well as help to provide a secure foundation for future postgraduate training.

Newer medical courses have replaced traditional teaching of the basic sciences with problem-based learning (PBL), where clinical scenarios are used as tools for students to identify relevant basic science concepts. Lectures and practical activities are structured around these clinical cases to attempt integration of basic science knowledge. In this current context, anatomy teaching and assessment has shifted away from anatomical dissection. Newer curricula utilize prosected specimens, models, computer-generated models and medical imaging to teach anatomy. Drake [1] in his classical publication addressed 4 main principles to guide curricular change and reform: first, medical education is a continuum. Second, integration of the basic sciences and clinical medicine should occur throughout the curriculum. Third, self-directed, lifelong learning should be emphasized, and fourth, it should emphasize active rather than passive learning and the development of lifelong learning and problem-solving skills while simultaneously decreasing lecture time. The recall knowledge is less emphasized.

Bergman et al. [2] have addressed, through a literature search, the debate on whether or not anatomical knowledge of today's medical students has been decreased. They found that despite the absence of empirical evidence for the existence of this problem, it is persistently emphasized by authors who lay the blame for it on one or more of the following factors:

1. Decrease in anatomy teaching time
2. Integrated curricula (problem-based learning or systems-based curricula)
3. Neglect of vertical integration of anatomy teaching
4. Anatomy is not taught in context
5. Decreased use of dissection as a teaching tool
6. The absence of a core anatomy curriculum
7. The way anatomical knowledge is assessed
8. Anatomy is taught by non-medically qualified teachers

Those factors, whether they contribute to the reduction in the anatomical knowledge of today's medical students or not, if such an assumption exists, are worth discussing in terms of their relevance towards the creation of an anatomical program that reflects the continuing nature of medical education. It will also provide with the opportunity to deepen students' understanding of anatomical structures and foster the development of life skills that will benefit them tremendously in the transition to graduate programs. In the subsequent sections, those factors along with the establishment of such a program will be discussed.

1. Less Time for Anatomy Education

Due largely to the explosion of information related to molecular biology, the introduction of new courses and concepts like professional skills, professionalism, ethics and attitudes, the basic science curriculum has become more crowded and the time for anatomy has become compressed to accommodate these new disciplines. Turney [3] addressed the issue of the decline of Anatomy as a subject. He reported that it is very difficult to assess objectively whether this reduction in anatomy teaching has been excessive and that the few studies that have been conducted suggest that the knowledge of the qualifying doctor is now below an acceptable level. He concluded that anatomy has suffered as a subject because it is regarded as didactic, traditional, overly factual and unable to adapt to modern educational methods which has made it an obvious target for those looking to reduce curriculum content and modernize the learning experience. Financial constrains; and changes in anatomy departments and surgical training have reduced the numbers of medically qualified anatomy teachers. This is within the context of an increase in numbers of medical students which puts more pressure on an overstretched system. Also, maintaining a dissecting room is becoming very costly.

There are a number of reports documented in the literature of increases in medical errors arising from inadequate knowledge of anatomy. These errors are frequently made at junior level and are associated with significant morbidity and mortality leading to a rise in litigation [4].

There have been a number of recent reviews describing the increase in litigation in surgical practice resulting from a lack of appropriate anatomical knowledge. In addition, it has often been observed in litigation cases that problems can arise consequent upon a practitioner's failure to explain to patients the full risks of a procedure because of their own lack of understanding and appreciation of the possible complications. Clearly, effective communication requires the appropriate knowledge and understanding [4].

2. Anatomy in Different Teaching/ Learning Modules

The unit or building block of anatomy is an anatomical structure or organ. Organs may be grouped according to a common function into *systems* or according to a common location into *regions*. An organ is therefore simultaneously the structural and functional unit of a body system as well as an occupant of a region [5].

2.1. Anatomy in Traditional Courses

The traditional learning model (also referred to as the lecture-based or faculty-centered/discipline-based curriculum) consists of faculty instructions, usually in a lecture and laboratory settings. Within traditional courses, anatomy programs typically had a 'regional' organization where learning tasks are sequenced in a linear fashion. The courses were department-based, and there was little integration across disciplines. As a result of information explosion, the usefulness of this approach began to be minimized with number of lectures and laboratories declining, and passive superficial learning dominating the scene. Students were required to memorize large quantities of facts, with little or no reference to clinical significance of the information learned. This model, in its basic format, was not an appropriate approach for the education of the future doctors [1].

2.2. Anatomy in Courses with Systemic Approaches

The systems-oriented learning model is based on a plan of how material is organized rather than how it is presented. It moves through each system in the body and usually involves the integration of several disciplines. The level of integration may vary, with some courses being included while others present their material in a discipline-based traditional format.

The presentation of material in this model is usually along the lines of the traditional approach, in which the lecture and laboratory are the main avenue for the dissemination of information. A purely systemic program is equally sequential and linear whereas it aids the viewing of relationships within the same type of structure; it deemphasizes important relationships between neighboring structures [5]. The strength of this approach is that it presents material in what should be a more logical fashion depending on the degree of integration between disciplines. However, the use of the traditional style for the presentation of material still fosters passive learning that is detail-oriented [1].

2.3. Anatomy in Problem-Based Learning Courses

In a problem-based learning model, the student is actively involved in identifying learning objectives and selecting personal learning resources. Its goal is to help the student develop a useful knowledge base, strong self-direction, life-long learning capabilities, and effective problem-solving skills that will be used in the clinical setting.

2.4. Advantages of Teaching Anatomy in PBL Systems

PBL courses are aimed to achieve horizontal (discipline-discipline) in addition to vertical (basic-clinical) integration. This has led to a systemic approach in teaching anatomical sciences. PBL curricula have the advantage of ensuring that students are engaged in a so-called spiral approach of learning, in which basic concepts are introduced at an early stage of the curriculum, and are revisited at greater depth at regular intervals across the entire curriculum.

The anatomy course spreads over the entire undergraduate curriculum rather than being delivered in a concentrated course in preclinical years, with the added benefit of introducing patient contact from the start of medical education.

2.5. Disadvantages of Teaching Anatomy in PBL Systems

In the past, clinical relevance did not play a major role in the design of the anatomy curriculum because curricula were discipline oriented. Currently, the disciplinary perspective could be becoming lost from PBL courses that have traded it for the prevailing clinical perspective. In many curricula, from being the major longitudinal traditional course, anatomy became fragmented with even the order of its fragments dictated to by the PBL tutorial cases. The regions of the body that did not fit in tended to be ignored. For example, the brain and spinal cord as the central nervous system were given appropriate emphasis in an integrated systems organization, whereas the surrounding head, neck and back were much more difficult to classify.

Some students may not function well in this type of self-directed learning situation. Furthermore, if knowledge gaps exist, some students will not fill in these gaps if left on their own [5]. The basic question that must be answered is: will this approach provide the best educational experience for the typical student at the institution?

3. How Should Anatomy Be Taught?

Of all the factors adversely affecting the teaching of anatomy, probably the most pressing is that of available time. This must stimulate us to focus on the purposes of our teaching methods and what students really need to know to become safe and effective clinicians [6].

3.1. The Ideal Medical Course Is Likely to Be a Hybrid Incorporating the Best of Traditional Courses and PBL Courses

Drake [1] suggested a two year program that integrates basic and clinical sciences as well as the traditional and PBL model in an integrated curriculum that takes the best aspects of each. The traditional program would provide lectures to outline basic concepts and laboratories to foster student discovery and group interaction. The problem-based learning format would be used in the discussion of weekly clinical cases to reinforce basic concepts. Finally, the systems-oriented model would provide the overall organizational plan for the presentation of material. The first year of the program is to orient the presentations toward normal structure and function and the second toward diseases.

As Drake puts it nicely [1], “for anatomists, the integrated model should be a very logical extension of what we have done in our courses for many years. We have used lectures to disseminate information, clarify basic concepts, and demonstrate the connection between basic knowledge and the clinical setting. The laboratory has always been an excellent setting where groups of students learn to work together as a team, working through and discussing problems as they proceed.

Additionally, many of our programs already have some type of clinical case presentations or discussions that foster self-directed learning at a minimal level and develop problem-solving skills. Anatomists have the programs that can fit into various curricular models. We now must demonstrate to the decision-makers that we, and our programs, are adaptable to whatever format is chosen" [1]. Both dissection and PBL tutorials, with learning tasks interpreting the anatomical basis of clinical phenomena have vital roles. The former links well with regional anatomy, while the latter with systemic anatomy.

3.2. Integrating Anatomy Horizontally and Vertically in the Syllabus

In the ideal new anatomy program, the disciplinary and clinical perspectives should be genuinely integrated. Historically, anatomy has been taught predominantly in the first undergraduate year. Although anatomical knowledge is assessed again in many specialist professional examinations, there is very limited exposure to anatomy teaching in later training. This is educationally unsound, as an excessive amount of seemingly irrelevant material in a curriculum encourages superficial learning. Given only one opportunity to teach anatomy, it is difficult to calculate the quantity of anatomical detail that should be included; should a course be preparing future generalists or specialists? A solution would be to integrate anatomy vertically into medical education so that students are exposed to anatomy teaching throughout undergraduate (preclinical and clinical), postgraduate and later professional training. This would provide relevant anatomy at an appropriate level of detail to the stage in training or career development. Tailoring theoretical learning to a particular specialty would reduce the amount of unnecessary theory learnt. [3]

In order to form a bridge between anatomy and the clinical subjects the study of human structure and function should occur in the context of clinical application based on the delivery of an integrated curriculum. In order to fully appreciate why certain facts about structure and function need to be studied, and to ensure reasonable retention of the knowledge, the delivery of anatomy should ideally be integrated with the clinically relevant details [5]. Furthermore, faculty should, as a goal, integrate the material delivered in macroscopic anatomy, clinical anatomy, histology, embryology and physiology (intra-discipline). Rather than simply tacking on applications to an already large syllabus, applications should be incorporated within the teaching program. Some of the learning needs may be already incorporated within PBL tutorials, while the remainder can be included in practical classes or ideally via independent learning utilizing suitable interactive multimedia [5]. Involvement of clinical specialists would give them the opportunity to shape the anatomy syllabus according to good clinical practice and advancing techniques, maintain their own knowledge (making them safer practitioners), and help to address the staff shortages in anatomy teaching. The criticisms of specialists about their juniors' lack of anatomical knowledge would be addressed directly and it should produce safer, more competent practitioners, less likely to make mistakes and incur litigation in the future [3].

3.3. Anatomy Is Best Taught and Learned in Context with Emphasis on Structure and Function in Health and Disease

The contexts that are commonly used in anatomy teaching can be roughly divided into four categories [2]:

(1) Clinical skills (physical examination)
(2) Pathology
(3) Radiology (e.g. interpreting X-rays, CT, ultrasound)
(4) Surgical procedures (e.g. Endoscopic/laparoscopic procedures).

There is an enormous potential in learning anatomy in the context of need. A curriculum in anatomy should provide the basis for physical examination of patients, interpretation of medical imaging and competence in basic procedures, including emergencies with room for flexibility of schools to utilize resources to best advantage. It is not sufficient to know about anatomical structures in theory but it is also vital to experience how normal structures appear from the variety of ways they can be accessed in living patients.

Sectional anatomy of the body at clinically important levels and in key planes forms the basis for interpreting computerized tomographic, magnetic resonance, and ultrasound images. Surface anatomy (including projections of underlying organs), together with functional anatomy (movements, actions and reflexes), form the basis for conducting the physical examination of a patient, as well as performing clinical, including emergency procedures.

There are many advances on the horizon, particularly involving imaging techniques both diagnostic (e.g. ultrasound) and interventional (e.g. angioplasty), as well as endoscopic and microsurgical procedures. A thorough knowledge of anatomy is now more important than ever, in order that the diagnostician does not see too little or too much [5].

Bergman et al. [7] in an article which originated from several studies investigating the knowledge of anatomy of students at eight Dutch medical schools reported that the studies showed that undergraduate students uniformly perceived deficiencies in their anatomical knowledge when they started clinical training regardless of their school's didactic approach. A study assessing students' actual knowledge of clinical anatomy revealed no relationship between students' knowledge and the school's didactic approach.

Test failure rates based on absolute standards set by different groups of experts were indicative of unsatisfactory levels of anatomical knowledge, although standards differed markedly between the groups of experts. When explored the differences in knowledge between students from different schools, they found that differences in the amount of time spent on anatomy varied with students' performance to some extent. Another factor that seemed to be associated with differences in anatomical knowledge was the frequency of their recurrence in the course of the curriculum. However, the authors noted that while clinical topics are regularly integrated into the early years of medical curricula, integrating basic sciences in the later curricular years was a tougher task. Their exploration of anatomy training grew even more interesting when they found the highest levels of anatomical knowledge among the students who were taught anatomy in a clinically meaningful context—such as patient problems, and diagnostic and therapeutic features. This was true for two schools with the best anatomy results in the comparison. At the time of the study, one of these schools offered a traditional and the other one a PBL curriculum, but they both taught anatomy in a clinical context.

They concluded that good test performance by students seems to be related to total teaching time for anatomy, teaching in clinical context, and revisiting anatomy topics in the course of the curriculum.

These factors appeared to outweigh the effects of disciplinary integration or whether the curriculum was problem-based or traditional.

3.4. Common Clinical Cases Should Drive Students Inquiry

Anatomy, like any other discipline, needs to be constantly revised and examined to demonstrate competency, especially when integrated into the clinical setting as students and even after graduation. In order to rationalize the potentially overwhelming workload, students require perspective into what is important for clinical competency, preferably in advance of trying to learn about a particular topic. Some researchers have asked students in different stages of the undergraduate curriculum or after finishing their studies whether they thought the contents of their anatomy course were of sufficient clinical relevance. While concurring to a certain degree, students also showed considerable differences of opinion on and between specific subjects (for references, I refer the reader to [2].

Rizzolo et al. [8] developed a series of pilot programs to explore how students learn anatomy and, in particular, how they combine instructional technology with more traditional classroom and laboratory-based learning. Their observations, experiences, and research had lead them to the following principles to guide the redesign of an anatomy course to make it more effective, less time consuming and more relevant to clinical practice:

1. The anatomy should be presented in multiple formats.
2. Common clinical cases should drive student inquiry. As an example, rather than dissect the facial nerve in the cadaver lab and then discuss why its relations are important, students were asked to prepare to perform a parotidectomy by investigating what structures will be placed at risk and the consequences of injuring them. The rationale is that dissection becomes more meaningful, because the students learn the anatomy in the context in which they will use it.
3. The pedagogy should be problem oriented. They reported that the radiology, hologram and web activities demonstrate that students crave practice and formative self-assessments. By extension, the dissection lab should be organized as a series of clinical problems that can be coordinated with radiology and web activities; they suggested.
4. The course should promote group process.
5. Self-assessment should be stressed over exam performance
6. Assessment of the course should continue into the clerkships

Based on the above principles; the group used common surgical and medical cases as the core of a revised anatomy course with anticipation that students will be more able to link what they have learned to new concepts presented in subsequent courses and the clerkships. In all, 15 disciplines recommended over 100 cases and surgical procedures that cover all regions of the body, and are common enough for a student to encounter similar cases during their clinical training.

A working group of clinical faculty, residents and students helped identify broad learning objectives, which it used to choose the best and more relevant clinical cases. Physicians then contributed the history and physicals, surgical notes, imaging studies and discharge summaries for relevant patients. All patient identifiers were removed to protect patient privacy. Idealized cases were developed from this raw data. Once the cases were developed, they refined learning objectives and began developing the dissection and conference exercises to achieve them.

During the course development process, they decided to begin the course in the same manner that a clinician begins with a patient – with the physical exam. They developed a lab that guides students over the anatomy of the body wall in a manner similar to a physical exam. This exercise introduces students to the skeletal system and surface anatomy. They then decided that the next lab should build on the first by asking students to perform clinical procedures that invade the body wall, rather than the time-consuming process of skinning the entire body and tearing apart muscle layers. Next to the clinical case approach, the most innovative aspect of the new course would be its reliance on formal small-group processes. To make up for the decrease in direct instruction by the faculty, students would prepare for the case-based labs under the guidance of a highly interactive, web-based dissector. The web-based dissector would guide their investigation of the anatomy that underlies the patient's presentation, physical exam, diagnostic imaging and surgical resolution. Students can link to web activities that lead them through the interpretation of plain films, contrast studies, MRIs and CTs.

In a similar experience, Drake [9] described the development of an Anatomy program at the Cleveland Clinic Lerner College of Medicine of Case Western Reserve University. Through consultation with surgery consultants, the case-directed anatomy program was developed. The program was effective, time efficient, and clinically relevant in the context of a non-lecture, problem-based, organ-systems-oriented curriculum. This approach uses clinical cases to introduce anatomical information that is reinforced using prosected cadavers and imaging.

The format of the approach involves three steps: (1) students preview a clinical case with clinical questions and learning objectives, (2) students acquire basic knowledge using textbooks and self-directed learning modules, and (3) students reinforce their basic knowledge in weekly case-directed anatomy sessions, which involve an interactive discussion of the clinical case followed by a laboratory. In the laboratory, students rotate around stations viewing prosected cadavers and imaging. The program continues after the first year, because it is longitudinal.

During the second year, students review anatomy in each organ-system course using the first-year format. Also, second-year students assist their fellows or residents prepare the prosections for first-year students and for their review of anatomy in the second year. This provides students with a dissection experience. During third-year clinical rotations, anatomy knowledge is reinforced, and several surgical anatomy electives were made available during the final year. In this way, anatomy is learned and reinforced throughout the medical school curriculum.

These were two excellent examples of a longitudinal approach to anatomy, where common clinical cases can guide the selection of content for an anatomy course that suitably prepares students for the rest of their medical training. Students learn anatomy because it is presented in the context of clinical situations and through the use of clinical examples. Thus, they see the importance of learning anatomy as they prepare for their future as physicians.

In Europe and North America interactions between anatomists and clinicians are facilitated by strong clinical anatomy associations. These associations perform several functions including the provision of a forum for the exchange of ideas on maintenance of standards of training in clinical professions, and a collective expertise that is able to comment on new initiatives, such as changes in medical training. Importantly, these associations also facilitate collaborative research by bringing biomedical scientists and clinicians together [10].

4. How Much Anatomy Should Be Taught?

4.1. The Need for a Core Curriculum

All those who design courses in medical schools, whether systems-based or integrated curricula, problem-based or scenario-led teaching, should ensure that their medical students learn in such a way that they become generally familiar with the core anatomy curriculum. This knowledge base is necessary for safe and effective examination and diagnosis of a patient. The syllabus defines the knowledge newly qualified medical graduates are expected to attain by the end of their course so that they can examine, diagnose and perform minor procedures safely and effectively. An important consequence of this is that such a standard of knowledge should be expected at the relevant level of examination, regardless of the teaching and learning methods used and the methods of assessment employed. This suggested minimum should serve as the benchmark for anatomical knowledge [4]. A minimum working knowledge should be that which allows an independent practitioner to practice safely, and to communicate with other medical professionals and patients effectively [3].

4.2. The Absence of a Core Curriculum and Its Effect on Anatomical Knowledge

In their literature search; Bergman and colleagues [2] reported that unfortunately, none of the articles they reviewed reported a study investigating the absence of a core curriculum and its effect on anatomical knowledge. The Royal College of Surgeons of England organized a meeting, in March 2007, to discuss the effects that reductions in anatomy teaching within the undergraduate curriculum were having upon the medical profession. One of the key consensus points to arise from that meeting was the need to define a core undergraduate syllabus in anatomy for all doctors [4].

Some authors have created a list of anatomical structures that students should know, in an attempt to define a core anatomy curriculum (For references, see Bergman et al. [2]). However, there is little agreement among authors on the content of such lists and authors do not define the depth of the required. If a core anatomy curriculum were in place, teachers, basic science and clinical educators, would know what students were expected to learn and it would help students to attain that knowledge [2].

4.3. Determining What Is Core Content of Anatomy

General anatomy (the core Anatomy) is conceptual and should be the anatomy that all health science students receive as their initial learning experience. Under the umbrella of General anatomy a group of specific anatomy components are held together [5]. General anatomy provides the foundation for specific anatomy in the clinical setting, including both of its systemic and regional theoretical perspectives as well as the practical [5].

The starting point of the process as described by Louw and colleagues [5]; should be to decide on the intended outcomes of the curriculum. Once they are determined, faculty will be able to work backwards, from the senior years to the junior years of the program, and set out what core information is required for the clinical settings and the basic and diagnostic sciences settings, following the concept of the spiral of learning. Faculty involved with senior and junior years (clinicians and basic scientists) have to be involved in the preparation of the curriculum, so that ideas can be shared and debated. This is an important way to ensure that only 'core' is delivered within a curriculum. It also brings about a healthy dialogue between what has traditionally been called the staff in the 'preclinical' and in the 'clinical' years which yields a greater commitment to delivering a program where there is more integration of subjects, both horizontally and vertically [5].

There have been a number of attempts to define a core syllabus in anatomy in the USA and Europe (Educational Affairs Committee, American Association of Anatomists, 1996 [11]; Griffioen et al. 1999. [12],; Kilroy and Driscoll, 2006 [13],; United states/ Canadian curriculum content in the Anatomical sciences; Summary of recommendations; a report of Working Groups of the Association of Anatomy, Cell Biology and Neurobiology Chairpersons, 1997; and General Plan Anatomy Objectives of The Teaching of Anatomy/Embryology in Medical Curricula in The Netherlands [14]. However, the syllabi defined by these documents are highly detailed either because they are aimed at a different level or targeted upon a more specialized audience for a more particular purpose.

Two other attempts to define a less detailed core of Anatomical knowledge worth reporting:

1) The work of Louw et al. [5]
   They outlined the general objectives of the general (core) Anatomy as well as specific objectives for general Anatomy with reference to the clinical application of each group of objectives. A comprehensive set of specific objectives and curriculum content (syllabus) was then drafted. Ideally a national or global core curriculum may ensure a minimum standard or at least provide a guide to assist anatomists in medical schools in constructing their own local adaptations, they recommended.
2) The work of the Anatomical Society of Great Britain and Ireland (ASGBI) through its Education Committee [4] has initiated a project to design an anatomical syllabus defining the minimum knowledge outcome expected in topographical anatomy, (excluding histology or embryology), in a graduating medical student in the UK and Ireland. This syllabus was drafted by the authors, the majority of whom are members of the Education Committee of Anatomical Society of Great Britain and Ireland. The authors are practicing teachers of anatomy with experience in a wide range of curricula, involved in curriculum design in their own institutions and with extensive cumulative experience of other curricula through external examining. The Education Committee of ASGBI includes members of the British Association of Clinical Anatomists. [4]. The syllabus is laid out in broad regional terms. It starts with a broad general statement which includes surface anatomy, the interpretation of standard clinical images, and the importance of the knowledge for understanding of common pathologies. This is followed by a more detailed specification of the topographical knowledge required to meet these general aims (The readers are referred to-A core syllabus in anatomy for medical students- Adding common sense to need to know;

prepared by the Education Committee of the Anatomical Society of Great Britain and Ireland [4].

4.4. Anatomical Components of Different Body Systems

Heylings [15], in a study to review the impact of Tomorrow's Doctors on anatomical teaching in the UK and Ireland; surveyed 21 anatomy departments. Twelve departments used systems-based curricula, five used problem-based curricula, and four used a traditional regional format. There were variable levels of integration between the anatomical disciplines and subjects outside anatomy. Overall, the various components of anatomy were all taught over 1 or 2 years (average 1.9 years for gross, 2.0 years for histology, 1.6 years for embryology and 1.5 years for neuroanatomy). These overall average times were raised slightly due to longer teaching time in the problem-based curricula, where they were taught over 5 years. Gross anatomy was taught using a regional approach in four of the traditional, one problem-based and in six of the systems-based curricula. One program had a dual approach using systems and regional approaches in the teaching of gross anatomy concurrently.

Not all curricula agreed on the suggested breakdown of anatomy into 10 systems. Many did not include the immune and nervous systems as separate systems. There was considerable variation with no obvious pattern in the order in which each of the systems could be approached.

However, by averaging all responses, the average curriculum commenced with the cardiovascular system and ended with the reproductive system. Those curricula designed around systems tended to agree more, with the cardiovascular, respiratory and musculoskeletal systems most likely to be taught first. The reproductive and endocrine systems were more likely to be taught at the end of the course, although in some cases the endocrine system was taught at various stages throughout the course. A few courses used a foundation module or term before the main systems teaching commenced, although it was not clear whether this was to accommodate an anatomical foundation or not. The problem-based curricula showed just as much variation, with cardiovascular, respiratory and endocrine systems most likely to be taught early and metabolism likely to be taught either first or last. Gross anatomy was fully integrated in the problem based curricula and was integrated to varying degrees in systems-based curricula, but little or no integration was identified within the traditional curricula.

NBME subject Exam's content outline for the subject of Basic Science Gross Anatomy and Embryology [16] presented the percentage of Anatomy and embryology contents in each of the 10 systems the exam tests. These exams are constructed to be appropriate for a broad range of curricular approaches. This distribution of the contents of Anatomical sciences to different body systems can be utilized as a guide while constructing the Anatomy curriculum. (http://www.nbme.org/Schools/Subject-Exams/Subjects/ basicsci_gross.html).

5. To Dissect or Not to Dissect; That Is the Question

Much debate has arisen about how to teach anatomy. This polarizes into those that favor dissection of human cadavers (traditionalists, predominantly surgeons and anatomists) and those that support newer teaching modalities (e.g. self-directed learning, problem-based learning [PBL], and computer-assisted learning).

Dissection has become synonymous with traditional courses and has come to be regarded as the contrast of problem-based learning (PBL). However, dissection would appear to be ideally suited to self-directed learning: students exploring a subject for themselves at their own pace, in a practical way and according to their own personal interests. It is a useful tool familiarizes students with the variation in anatomy and to obtain an appreciation of fully exposed structures that cannot be seen through the window of an operation but might be damaged unintentionally during surgeries [3].

Perhaps most significantly, students have a high regard for dissection as a learning resource in the anatomy course and many other learning outcomes have been identified by students.76% of anatomy courses in the UK and Ireland involve dissection, irrespective of curricula design [15].

Kerby et al. [17] concluded that the overall preferred teaching method of anatomy to achieve the best learning outcomes is with human cadaveric dissection, as perceived by medical students. This opinion of medical students mirrors that of professional anatomists as reported by Patel and Moxham [18].

Professor Harold Ellis [19]; has clearly stated the benefits of dissection as follows:

- teaches the basic language of medicine
- teaches some manual dexterity
- introduces an understanding of three-dimensional anatomy
- introduces the concept of biological variation
- introduces some common pathological conditions
- introduces students to team working
- introduces communication skills between individual students and groups
- acclimatizes students to the reality of death and teaches respect for the body
- is a prime example of 'self-directed learning'

There was a consensus on the importance of anatomy in medical curricula and most anatomists were receptive to change. The most important finding was that professional anatomists, even modernists, supported human cadaveric anatomy as a superior tool to achieve the aims of the course. [15]

6. Future of Anatomy in Outcome- Based Medical Education

One of the major pedagogical movements in recent years has been toward outcomes-based medical education, which focuses on the outcomes displayed by trainees rather than didactic content, delivery methods, duration of training programs, and other inputs [20].

The competency model evolves over the continuum of medical education, with different expectations for individuals as they move from being college undergraduates to medical students, residents, and practitioners. The gap between education and practice can be bridged by competency-based education [21]. Medical schools and residency programs will need to work together to develop a blue print for seamlessly linking undergraduate and graduate training [22].

Learning outcomes for individual curricula or programs are often situated within broader outcome-based competency frameworks, particularly at the postgraduate level. In the United States, there are six core competencies required of residents by the Accreditation Council for Graduate Medical Education (ACGME), which set standards for postgraduate training and increasingly influence continuing medical education for maintenance of certification in practice. The ACGME competencies include Patient Care, Medical Knowledge, Practice-Based Learning and Improvement, Interpersonal and Communication Skills, Professionalism, and Systems-Based Practice.

The Royal College of Physicians and Surgeons of Canada established the CanMEDS competency framework. It defines roles of the physician as Medical Expert, Professional, Scholar, Advocate, Communicator, Collaborator, and Manager (for references, please see [20]). The Association of American Medical Colleges (AAMC) Medical Student Objectives Project Report I; concluded that physicianship could be broadly categorized under the competencies of knowledge, skill, altruism, and duty [23]. The Institute of Medicine in 2001 made broad recommendations that medical education should emphasize patient care delivery in multidisciplinary teams, evidence-based practice and continuous quality improvement, the effective use of information technology, and lifelong-learning skills (Institute of Medicine, Committee on Quality of Health Care in America, 2001 [24]. Although there has been some initial resistance, these competency-based frameworks are now being widely adopted and appear to be generally applicable to clinical training programs.

In designing appropriate outcomes for an Anatomy curriculum, it may be desirable not only to integrate anatomy with clinical practice but also to include objectives that relate to a wide range of competencies. A model to consider; is the recent curricular redesign at Mayo Medical School which presented the opportunity to place greater emphasis on the six core competencies in the first 2 years of the medical curriculum and to transition basic science courses to outcomes-based models. Gross anatomy is one of several basic science courses from the old curriculum that was completely revised. The ACGME core competencies drove the process of integrating first-year gross anatomy, radiology, and embryology into one course. As a result, a 6-week (120-contact-hour) hybrid gross anatomy, radiology, and embryology didactic block entitled ''Human Structure, 'comprised of more than 20 educational interventions that each targets one or more core competency, was implemented [25].

7. Assessment

Assessment systems in anatomy are dealt with in other chapters of this book (14 through 16). In brief you assess students the way you teach them. Assessment systems must be aligned with the teaching methods and the stated learning objectives or outcomes of the curriculum/course. This needs to be clearly tabulated for all teachers and examiners, as well as students, so that everyone is equally involved in the process (blue printing). By seeing how each objective is addressed by a particular learning activity and related to the assessment (mapping objectives to assessment tools), students are in an informed position to decide on the best way and how much effort is needed to achieve the objectives so that they can perform well when they are assessed. The amount of material to be learned in a typical basic science course is overwhelming, and students must make choices of what to study. As a result, medical students are highly motivated and academically strong, they want to perform well on tests, and they look to tests for guidance on what material is worth studying. To the extent that tests are congruent with major course goals, the tests will appropriately communicate what should be learnt [2]. The mapping can also be extended horizontally and vertically to other subjects with integration or at least coordination of the program regarding sequence of major topics and, most importantly, the timing of assessments [5].

In an interesting study, Craig et al. [26] surveyed 19 Australian and New Zealand medical schools, examining the time allocation, content, delivery and assessment of anatomy for the 2008 academic year. The results were as follows; Multiple-choice questions were the most commonly used means of assessing gross anatomy (16 of the 19 medical schools). Modified essay questions were used by 12 medical schools, and extended matching questions were used by six schools. Nine of the 19 medical schools utilized practical-based identification for assessment of gross anatomy. Nine of the 19 medical schools required students to complete a specific anatomy assessment. The majority of medical schools (10 of 19) integrated anatomical assessment into applied cases with other basic sciences in problem-based assessments. Because anatomy was integrated with other content for assessments in most courses, there was no minimum level of achievement set for gross anatomy in 11 of the 19 courses, only for the overall integrated paper. In such courses, students can perform very poorly in gross anatomy but do well in other disciplines and still pass the course (as specifically commented by several universities). The remaining universities (8 of 11) required a pass grade (50%) or higher in gross anatomy to progress within the course.

In general; the aim of all assessment techniques is to place the basic science of anatomy into a meaningful clinical context.

Conclusion

There are substantial changes underway in medical education, especially the competency-based model. Curriculum reform at individual medical schools can be slow and long process. The key is to incorporate knowledge of the basic sciences, anatomy included, into the clinical activities that comprise the competencies mandated by undergraduate and postgraduate training. The diversity of anatomy teaching at different institutions remains a major issue. While differences can never be eliminated, the implementation of a core anatomy curriculum or framework would be very beneficial to ensure that minimal anatomical competency is reached by the graduate at graduation.

The relevance and purpose of what we teach should be clearly established. Knowledge of anatomy forms the basis for much clinical examination and a necessary prerequisite for many medical procedures and may form the basis from which to begin to define core anatomy. Anatomy should not be viewed as an isolated preclinical discipline and it should be horizontally and vertically integrated throughout the undergraduate medical curriculum and beyond. In times of changing medical curricula and of increasing pressure for time from other disciplines it is important that anatomists continue to be involved with all stages of medical education. Co-operation between anatomists and clinicians is essential to determine what is taught, by whom, and when.

In Summary; Anatomy should remain a principal component of medical education, with dissection as its core teaching method. Anatomy must be linked with modern approaches to medical education and it should be integrated, continuous and clinically relevant. It is critical that clinicians engage in the teaching of anatomy in the clinical environment and they must be adequately resourced to do so. Further collaboration between clinicians and anatomists should strengthen a mutually beneficial relationship.

REFERENCES

[1] Drake, R.L., Anatomy education in a changing medical curriculum. *Anat. Rec.,* 1998. 253(1): p. 28-31.

[2] Bergman, E.M., C.P. van der Vleuten, and A.J. Scherpbier, Why don't they know enough about anatomy? A narrative review. *Med. Teach.,* 2011. 33(5): p. 403-9.

[3] Turney, B.W., Anatomy in a modern medical curriculum. *Ann. R Coll. Surg. Engl.,* 2007. 89(2): p. 104-7.

[4] S. McHanwell, M.A., D.C. Davies, R. Dyball, J. Morris, C. Ockleford, S. Standring, I. Parkin, S. Whiten, J. Wilton, A core syllabus in anatomy for medical students-adding common sense to need to know. Rev. from *Eur. J. Anat.,* 2007. 11(Supplement 1): p. 3-18.

[5] Louw, G., N. Eizenberg, and S.W. Carmichael, The place of anatomy in medical education: AMEE Guide no 41. *Med. Teach.,* 2009. 31(5): p. 373-86.

[6] Abu-Hijleh, M.F., The place of anatomy in medical education: Guide Supplement 41.1--Viewpoint. *Med. Teach.,* 2010. 32(7): p. 601-3.

[7] Bergman, E.M., et al., How much anatomy is enough? *Anat. Sci. Educ.,* 2008. 1(4): p. 184-8.

[8] Rizzolo, L.J., et al., Design principles for developing an efficient clinical anatomy course. *Med. Teach.,* 2006. 28(2): p. 142-51.

[9] Drake, R.L., A unique, innovative, and clinically oriented approach to anatomy education. *Acad. Med.,* 2007. 82(5): p. 475-8.

[10] Nicholson, H., The future of anatomy: where do we go from here? *ANZ J. Surg.,* 2005. 75(3): p. 95.

[11] EDUCATIONAL AFFAIRS COMMITTEE, A.A.O.C.A., A clinical anatomy curriculum for the medical student of the 21st century: gross anatomy. *Clin. Anat.,* 1996. 9(2): p. 71-99.

[12] Griffioen, FMM., Drukker, J., Hoogland PVJM., Godschalk M., General plan anatomy. Objectives of the teaching of anatomy/embryology in medical curricula in The Netherlands. *Eur. J. Morphol.,* 1999. 37: p. 228-325.

[13] Kilroy, D. and P. Driscoll, Determination of required anatomical knowledge for clinical practice in emergency medicine: national curriculum planning using a modified Delphi technique. *Emerg. Med. J.,* 2006. 23(9): p. 693-6.

[14] General plan anatomy. Objectives of the teaching of anatomy/embryology in medical curricula in The Netherlands. Appendix 2. *Eur. J. Morphol.,* 2000. 38(5): p. 355-68.

[15] Heylings, D.J., Anatomy 1999-2000: the curriculum, who teaches it and how? *Med. Educ.,* 2002. 36(8): p. 702-10.

[16] NBME subject Exam's content outline for the subject of Basic Science Gross Anatomy and Embryology http://www.nbme.org/Schools/Subject-Exams/Subjects/basicsci_gross.html; (Last accessed on December 23, 2012).

[17] Kerby, J., Z.N. Shukur, and J. Shalhoub, The relationships between learning outcomes and methods of teaching anatomy as perceived by medical students. *Clin. Anat.,* 2011. 24(4): p. 489-97.

[18] Patel, K.M. and B.J. Moxham, The relationships between learning outcomes and methods of teaching anatomy as perceived by professional anatomists. *Clin. Anat.,* 2008. 21(2): p. 182-9.

[19] Ellis, H., Teaching in the dissecting room. *Clin. Anat.,* 2001. 14(2): p. 149-51.

[20] Hazelton, L., Changing concepts of neuroanatomy teaching in medical education. *Teach. Learn Med.*, 2011. 23(4): p. 359-64.

[21] Tilley, D.S., et al., Promoting clinical competence: using scaffolded instruction for practice-based learning. *J. Prof. Nurs.,* 2007. 23(5): p. 285-9.

[22] Harden, R.M., Learning outcomes as a tool to assess progression. *Med. Teach.,* 2007. 29(7): p. 678-82.

[23] Association of American Medical Colleges (AAMC). 1998. Medical SchoolObjectives Project (MSOP) Report I. Learning objectives for medicalstudent education: Guidelines for medical schools. Washington, DC.pp. 1–13. Available from: https://services.aamc.org/publications/showfile.cfm?file¼version87.pdfandprd_id¼198andprv_id¼239andpdf_id¼87.

[24] Institute of Medicine, Committee on Quality of Health Care in America(IOM). 2001. *Crossing the quality chasm: A new health system for the21st century.* 1st ed. Washington, DC: National Academy Press. p 337.

[25] Gregory, J.K., et al., Restructuring a basic science course for core competencies: an example from anatomy teaching. *Med. Teach.*, 2009. 31(9): p. 855-61.

[26] Craig, S., et al., Review of anatomy education in Australian and New Zealand medical schools. *ANZ J. Surg.,* 2010. 80(4): p. 212-6.

In: Education in Anatomical Sciences
Editor: Paul Ganguly
ISBN: 978-1-62618-488-6

Chapter 4

How Do We Teach Anatomy? Use of Team-Based Learning Strategy

Cheryl Melovitz-Vasan, David O. DeFouw and Nagaswami S. Vasan*
Department of Cell Biology and Molecular Medicine,
New Jersey Medical School, Newark, New Jersey, US

Abstract

Knowledge of anatomy forms the keystone of medical education, but to teach anatomy has become a challenge due to various reasons and at the same time presented an opportunity for innovation. The challenges resulted in an innovative approach that meets the expected criteria of student engagement in their learning, teaching efficiency and effectiveness, and economical delivery. The pedagogical model- 'Team-Based Learning' (TBL) meets the need of both administration and "millennial Learners". Furthermore, TBL combines elements of adult learning principles and other self-directed experiential learning models in a distinctive way to teach anatomy, which is theoretically complex and content intense. Readers should note several key components in the TBL strategy: first, teachers clearly identify content to be learned by setting learning topics and selecting resources for self-directed preparation. Secondly, instructors provide instruction for the learners who come prepared for team discussion of the learning topics. The sessions also include individual and group "readiness assurance tests". Thirdly, learners tend to be motivated to prepare for class and to perform well within groups. Such preparation is based on self-directed out-of-class study either alone or in teams. In TBL, the axiom is 'assignment drives group process and learning'. We find that when an assignment is simple and can be done easily by individual effort there is no team interaction. When the assignment is complex and truly requires the group to work together, there is more group interaction, enriched discussion and greater team satisfaction. We further observed that the students were more engaged in their learning as the teams facilitated active learning as well as peer teaching. To evaluate the possible effect of TBL on student performance, we compared the departmental exams and NBME subject exam scores between the traditional curriculum and TBL curriculum. We also

---

* E-mail: vasanns@umdnj.edu.

compared students' performances on the NBME exams in other first year required courses, and students' performances in anatomy on the licensing exam. This chapter focuses on illustrating how TBL foster team work, trust, mutual respect, interdependence and attitude towards learning for advance academic success. Medical education should be an active learning process that engages students to work in teams. It is a skill that is increasingly becoming vital and emphasized interprofessional education and team building.

INTRODUCTION

Anatomy is the cornerstone of medical education and anatomical knowledge is the foundation for training future doctors. But given the current climate of time constraint, increased class size, cost containment and the student population ("Millennial Learners") what are the best and most effective options available for teaching anatomy? The traditional model of a lecture-based faculty-centered approach has been the mainstay in many medical schools; however its reliance on passive learning has been challenged. The problem-based teaching/learning model, which is based on adult learning theory, has its own shortcomings. It is faculty intensive, requiring many facilitators and not all students learn well in this type of self-directed learning. Furthermore, if knowledge gaps exist, some students will not fill these gaps on their own and this has serious consequences. An appropriate model should include a brief outline of basic concepts that facilitates self-directed learning, and a focused study guide with opportunities to work in interactive small groups and frequent and timely feed-back. Small group interactions also provide opportunities to regularly apply learned knowledge to solve clinical cases. Furthermore, 21st century student approaches to learning demand a new way of teaching that is efficient, engaged, effective and economical, which paradoxically coincides with expectations of the administration. As the newest pedagogical strategy, team-based leaning (TBL) importantly addresses both student's expectations and administrative demands (Vasan et al., 2011).

TBL combines various elements of adult learning theories, and the curriculum model CASE- Cooperative, Active, Self-directed and Experiential Learning (University of Saskatchewan, College of Medicine) in a new and unique way to teach anatomy which is theoretically multifaceted and content dense.

TBL was first developed in the 1970s by Professor Larry Michaelsen (Michaelsen et al., 2004) who implemented it for large-classes and changed lecture time into a small group interactive learning format. In place of lectures he provided assignments to prepare for class room discussion and problem solving (Michaelsen et al., 2004).

Student engagement characterizes TBL. The concept of TBL focuses on learning defined by pre-class preparation and subsequent in-class team discussion. In the absence of anatomy lectures, a series of organized learning activities ("content specific discussion topics" for team encounters) were provided to help students build baseline facts into a framework of conceptual interpretation and understanding (Vasan and DeFouw, 2005; Vasan et al., 2008; 2009; 2011; 2012). Faculty organizes the learning material to enable students to accept ownership of their own and their peers' learning thus creating a powerful learning team. TBL is based on individual accountability, insures rigorous team interaction to accomplish both competence (what individuals know or are able to do in terms of knowledge, skills and attitude) and capability (extent to which individuals can adapt to generate new knowledge and continue to improve performance). Michaelsen et al. (2008) put forth four principles of TBL-properly formed and managed teams, including appropriateness of assignments, student's accountability for the quality of individual and group work, and frequent and timely feedback that promote both team learning and development.

Classical Approach to TBL (Michaelsen et al., 2004)

Members of each team are randomly selected. Course structure involves concept oriented brief introduction in a large group format. It has 3 phases, where phase 1 involves out of class preparation, phase two, in-class Individual Readiness Assurance Tests (IRAT) and Group Readiness Assurance Tests (GRAT), and phase 3 involves application exercises that consist of cases in the form of vignettes with a subset of questions related to the case. In phase 3, all the teams announced answers for each question simultaneously to facilitate inter-team discussion.

Modified TBL (Vasan et al., 2008)

Based upon the diversity and demographic background of the class, we assign individuals to ensure well-balanced teams. This guarantees the same level of resources and opportunities for each group to develop into a "learning team" with cohesiveness and understanding. We create six member teams. Modification to the classical TBL is based on the needs and culture of our school. It is a common practice to use this pedagogical strategy to fit course structure, curricular time and other local factors.

Modified phases as described in Figure 1.

Phase 1: We assigned readings from required texts and created "content specific discussion topics" for team encounters (referred as 'discussion topics') based on the anatomy text and cadaver dissections.

Phase 2: Team encounters started with the IRAT (10 minutes) which is immediately scored by the staff and the results made available to the course coordinators. Immediate assessment of quiz performance by individual students allowed us to monitor each student's level of preparation and provide them appropriate feedback when warranted. IRATs are MCQ that are case based. Following the IRAT, teams discussed the assigned 'learning topics' (90-120 minutes) to solidify various course concepts for deeper understanding. Faculty provides clarification when required. The sessions are concluded by teams taking the same MCQ quiz, now defined as GRAT (20 minutes).

Phase 3: Inter-team case discussion (120 minutes) - Recently, we have introduced 'cases for team discussion' at the end of each module summing up the content. Teams discuss the cases individually, and then present it for inter-team discussion. These sessions are moderated by faculty who also use probing questions to expand student knowledge, clarify misinterpretations and address any knowledge gaps. This recent addition is the most popular part of the TBL since it provides better understanding of the content and helps the student in getting ready for the phase 4.

Phase 4: These are 'high-stake' exams that encompass the module content. Questions use clinical vignettes and are taken individually (IRAT) and as a group (GRAT).

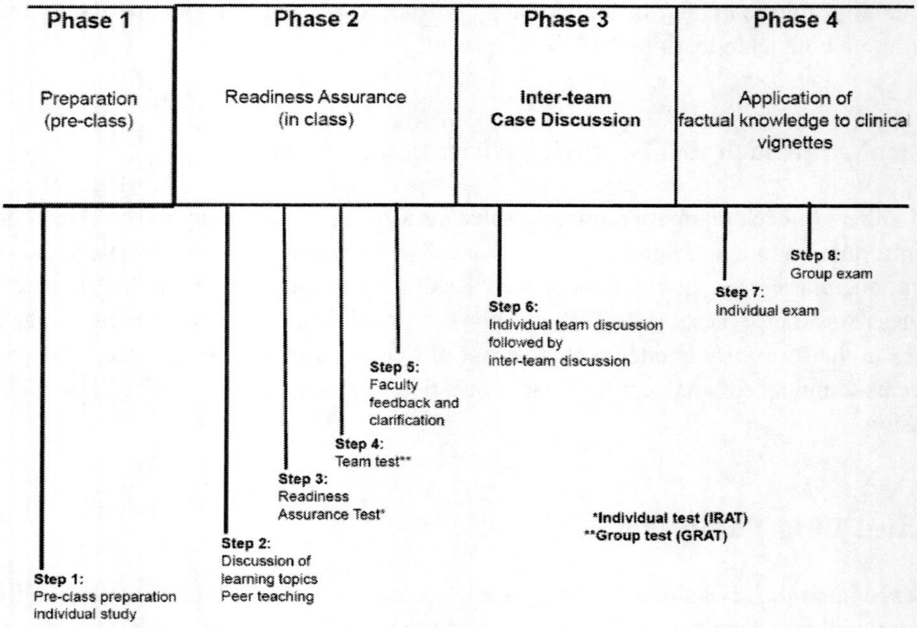

Figure 1. Four phases of TBL and the steps involved are shown. As described in the methods, the steps are different from the classical TBL originally suggested.

Concerns About Covering the Course Content

In a course like anatomy, with a vast amount of content traditionally taught by a team of research oriented basic scientists, many barriers existed against educational innovations. Lectures are economical compared to small group teaching and a new approach needs a considerable time investment. The decoy against change was a 'concern about covering the course content'. However, there are examples of content heavy science courses successfully utilizing TBL (Dinan, 2004; Nieder et al., 2005). We were convinced that the anatomy course content could be adapted to our TBL approach as described below.

To Provide Lectures or Reading Assignments and Learning Topics

At our school, seasoned teachers taught traditional anatomy with many lectures on details of anatomical structures and function. Irrespective of the lecture quality, we observed that student attendance at lectures had reduced considerably, and some students were there for mere physical presence or guilt. Students were successful in anatomy examinations irrespective of their lecture attendance. Thus, the question begging for an answer asked whether lectures were superfluous for learning anatomy: We began to question the effectiveness of lectures for deeper student learning, critical thinking and long-term memory. We find a lecture is where the notes of the professor are transferred to the notes of the student without passing through the brain. But, it was hard to give up lectures for those who were used to being a "the sage on the stage" to become a "guide on the side" (Herreid, 1998). Starting in academic year 2004, in place of lectures we introduced 'content specific learning topics' (learning topics) for team encounters. These were for students' out-of-class preparation and focused more on students taking the initiative to learn and to teach their peers (principles of adult learning, cooperative learning and engaged learning). The learning topics focused on materials relevant to clinically correlated anatomy that allowed them to solve clinical problems through critical thinking. To ensure that all the students were properly prepared, we began all team encounters with an ungraded IRAT. The IRAT scores were tabulated immediately and students were given appropriate feedback. What we observed was astonishing. There was 100% class attendance. The transformation of students' excitement and enthusiasm was astonishing as was their preparedness for team discussions. We now had an answer: students learn anatomy not from lectures but through focused learning topics and an opportunity to apply their knowledge for clinical reasoning and team discussion. We also observed that the TBL encounters were intense with students coherently verbalizing their thoughts to explain discussion topics to each other. In most cases, students were not inhibited to ask for peer explanations of difficult concepts. The students learned more than specific anatomical knowledge and used many resources to expand their clinical relevance. The atmosphere was positive, built student trust, improved communication skills and helped to develop good interpersonal relationships.

How We Structured Team-Based Learning

The time commitment for putting together a TBL curriculum is the same effort that is required to devise a new course or curriculum. However, if the course is already established in lecture format (like our human anatomy), the principal requirement is converting the lectures into reading assignments and 'learning topics'. This is a substantial undertaking, which needs enormous time investment and is not to be taken lightly, since the success of TBL very much depends on it. Since medical human anatomy is a vast subject, we made sure that the reading assignments and 'learning topics' were unambiguous and focused on course goals and objectives. Providing focused 'learning topics' assured students that if they had difficulty with any items, there would be help from the group during team discussions. The reading assignments included lists of pages from the text to study, points of clinical relevance, concepts to be developed and resources to use. This offers students a clear idea of what knowledge is expected and what effort is needed to be successful in the course. Well organized learning activities are vital to help students build baseline facts into a framework of conceptual interpretation and understanding (Vasan and DeFouw, 2005; Vasan, DeFouw, and Holland, 2008; Vasan, DeFouw, and Compton, 2009, 2011). We maintained transparency and communicated to the students that there was no hidden curriculum.

Course Structure: The Human Anatomy course, which is offered in the fall semester, is divided into three units—thorax, back, and upper extremity (6 weeks); head and neck (5 weeks); and abdomen, pelvis, perineum, and lower extremity (8 weeks). Approximately 55-60% of course time is spent on cadaver dissection in small groups (four students per cadaver); 30-33% course time involves the TBL sessions and the remainder includes embryology and clinical correlation lectures. Faculty coverage of the laboratory (40–43 tables) is as follows: 10–11 dissection tables (40–44 students) are covered by a single faculty member, while the course coordinators rotate among all the dissection tables. For TBL, the students were assigned to teams of six, and 18-24 team encounters (team discussions) occurred during the semester. TBL involves four phases (Figure 1). Each TBL session was scheduled for 2 hours and often teams continued their discussion longer. All the TBL sessions were monitored full time by the faculty continuously rotating through the teams during their discussion. Clinical correlation lectures explained the anatomical basis of various clinical conditions (e.g., Horner's syndrome; Bells palsy; Pancoast syndrome etc.).

Student Assessment: The students take three-graded MCQ unit examinations (see course structure above), which are based solely on clinical vignettes. The examinations are taken both individually (scantron® scored) and by teams (IF-AT® form scored). Teams are also allowed to challenge any examination answers, and if a challenge is accepted, the result is applied to the entire class (unlike the classical TBL where only the challenging teams received credit). The comprehensive final examination consists of the anatomy and embryology National Board of Medical Examiners (NBME) Subject Examination and is taken individually only. NBME Subject Examinations for various subjects are made available to medical schools for a fee, and are administered according to their guidelines. The NBME scores the examinations and the grades are sent electronically to the Office of Education. The course grading rubric for student assessment is as follows: 40% from the individual unit exams, 15% from the individual NBME final, 30% from the individual practical exams and 15% from the team unit exams. Thus, 85% of the course grade for each student reflects individual achievement and 10% denotes a combination of team scores and peer evaluation.

Prior to introducing TBL, the course content was delivered via lectures that focused on factual anatomy without clinical correlation. Review sessions were typically offered prior to the course examinations. Assessment was K type MCQ that required recall of factual anatomy or embryology without clinical applications and also included true/false questions, fill in the blanks etc.

Evaluation of TBL

To evaluate the possible effect of TBL on student performance we compared the unit exams and NBME subject exam scores between the traditional curriculum and TBL curriculum (Table 1, Vasan et al., 2008).

Data were collected in Microsoft Excel and transformed into SAS data sets (SAS, Cary, NC) for analysis. Two-tailed t-tests were performed at alpha=0.01, to confirm that changes in scores were significant beyond chance fluctuations. A sign test was performed to confirm significant directionality of trends. We also compared the students' performances on the NBME exams in other first year required courses (Figure 2, Vasan et al., 2008). Furthermore, we compared students' performances in anatomy on the licensing exam (Figure 3, Vasan and DeFouw, 2012).

Table 1. Comparison of class averages in the course and NBME Subject examinations. Traditional curriculum refers to lecture based teaching. Scores for 2004 are not included since it was the year we piloted TBL. Scores for TBL curriculum were statistically different (p <0.05 (using two-tailed t-test) from the traditional curriculum score (2003) for each unit and for NBME examination

CLASS EXAM PERFORMANCES (AVERAGE %)		
CURRICULUM	UNIT exams	NBME
Traditional (2003)	73.3	64.0
TBL		
2005	84.6	72.0
CLASS EXAM PERFORMANCES (AVERAGE %)		
CURRICULUM	UNIT exams	NBME
2006	85.6	78.0
2007	85.0	79.0
2008	83.0	76.0
2009	83.3	80.0
2010	81.4	79.0

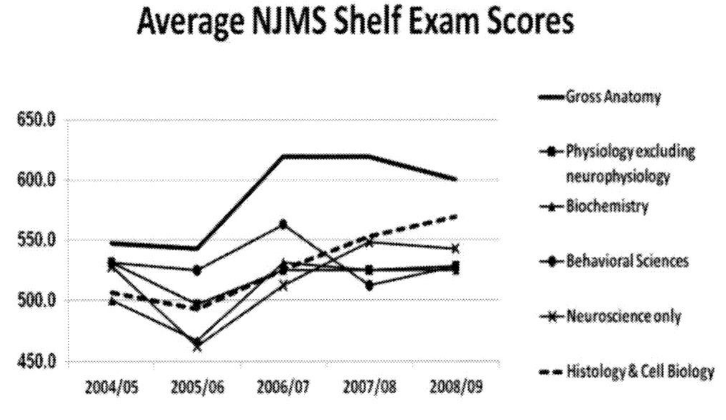

Figure 2. Comparison of student performance in the NBME Subject Examination. The national average score is 70. Other than anatomy all the others are lecture-based traditional courses. In anatomy, 2004 was the year TBL was piloted and 2005-2006 the years when modified TBL was fully implemented. Scores from other courses after 2009 were not made available.

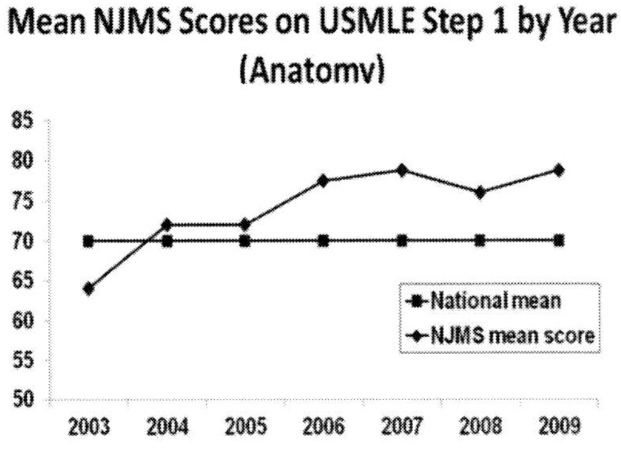

Figure 3. Student performance in USMLE Step I licensing exam in anatomy compared with national mean (provide by the NJMS office of education).

Outcomes

We piloted TBL in the anatomy curriculum in year 2004, and this allowed us to reflect and refine the process to develop the modified TBL that has been in place since 2005. Hence, students' performances during 2004 year are presented but not included for statistical analysis. The traditional lecture based curriculum offered in 2003 served as the basis for comparison. The class average GPA and MCAT scores for years 2003 were 3.47 and 29.6, and for years 2005-2010 varied between 3.51-3.64 and 29.9-31.6. This indicates that the entering classes included in the study are of comparable qualifications. The differences in either GPA or MCAT scores are not substantial and were not in any case subjected to statistical analysis.

Performances in Unit Exams and NBME Subject Exam

In Table 1, we compared the class performances from a lecture-based traditional curriculum (2003) and the years we implemented the modified TBL curriculum (2005-2010). In the unit exams, the class averages in year 2003 ranged from 70 to 77. There was a modest improvement in 2004 (averages ranged from 77 to 81) when we piloted the TBL concept (not included in the table). Importantly, subsequent to implementation of the modified TBL curriculum in 2005, the class averages, which ranged from 81 to 86, were significantly greater ($p< 0.01$) than those from year 2003. We further observed that individual students or teams who consistently performed well in the IRAT/GRAT performed well as individuals or as teams in the course exams.

For the comprehensive final, we administered the NBME gross anatomy and embryology subject exam. The individual scores in the NBME exam contributed to 15% of the course grade. Table 1 shows the class average scores in the last five years. The class averages in the years when TBL was implemented are significantly higher than the previous years when the course followed a lecture-based curriculum. For each unit, and for the NBME exam, differences in results were statistically significant when the TBL curriculum was compared to the traditional curriculum ($p<0.01$ in each case, using two-tailed t-test). Comparison of students' performances in the NBME subject exams in all the first year courses clearly indicate the advantages of TBL over the lecture based traditional curriculum followed by the other courses (Figure 2). Similarly, students performed exceedingly well in anatomy in the licensing exam (Figure 3) when compared to all the other preclinical (year 1 and 2) courses.

Studies have shown that TBL increased learner's engagement and preparedness, improved problem solving, communication and teamwork skills and knowledge outcomes (Thompson et al., 2007; and references therein). However, a systematic analysis of student experience or perception of TBL strategies has not been performed. Hence, we created a questionnaire to evaluate student perceptions of their TBL experiences using a questionnaire survey that elicited not only perceptions of pedagogy but also how perceptions may be related to learning inventory and process (Vasan et al., 2009). In brief, the questions probed student's preparation for team discussions, usefulness of learning issues for acquiring knowledge, importance of group discussion for deeper understanding, and attitudes about team behavior and mutual respect during team discussions.

In addition to questions related to perceptions of TBL, students were asked to provide the grade that they anticipated receiving in the course prior to administration of the final exam (Honors >90%; High pass 80-90%; Pass 70-80% and Fail <70%). We then used these grading categories to compare student responses on the two scales that were created using analysis of variance (ANOVA) with Bonferroni corrections to control for multiple testing effects. Results were considered significant if p<0.05.

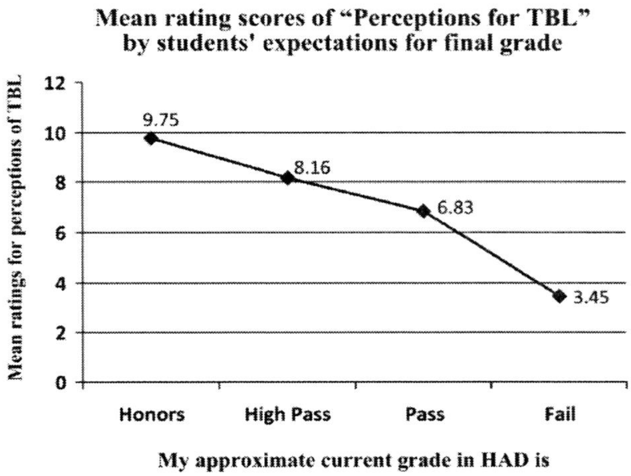

Figure 4. Mean ratings of students who expected a grade of Honors, High Pass, Pass, and Fail, respectively. One-way analysis of variance indicated there was no significant overall difference in mean ratings (p=0.084).

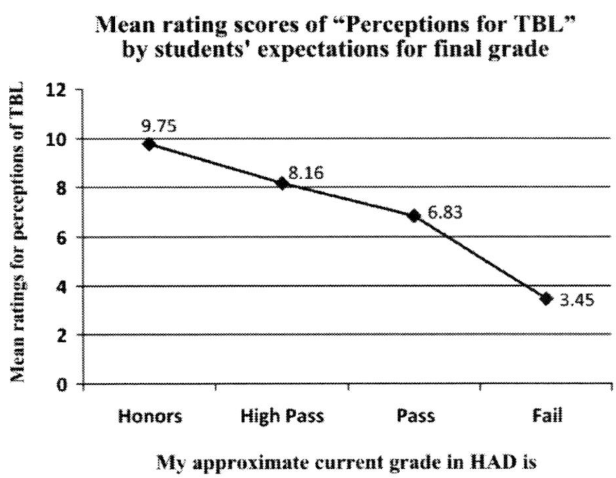

Figure 5. Mean ratings of students who expected a grade of Honors, High Pass, Pass, and Fail, respectively. One-way analysis of variance indicated a significant overall difference in mean ratings (p=0. 003).

A principal components factor analysis with varimax rotation identified two 8-item factors: "perceptions of TBL" and perceptions of "Teamwork". Internal consistency for each was high [Cronbach's alpha = 0.908 (Preference for TBL); 0.884 (Teamwork)]. Results of a one-way analysis of variance between Honors/High Pass/Pass/Fail students indicated that,

Honors (n=73) tended to rate "Perceptions of TBL" higher than Pass (n=54) [mean difference = 2.92; 95% CI (0.05, 5.79)], and also higher than Fail (n=11) [mean difference = 6.30; 95% CI (1.13, 11.47)] (Figure 4). However, each had overall positive ratings. No difference was noted between mean ratings of "teamwork", which were also, overall, positive (Figure 5). We conclude that, medical students view TBL favorably irrespective of their grades. While higher performing students tended to perceive TBL more favorably than lower performing students, overall, there was strong positive support for TBL. Furthermore, a large majority of students expressed positive feelings about teamwork.

Final Analysis

We consider our approach to TBL to be unique in that we replaced anatomy lectures with learning topics that enabled students to work individually and as a team in learning anatomy. We were also able to cover the course material in spite of reduction in curriculum time. Compared to passive learning associated with traditional lectures, team interactions allow more active student participation that fosters both activation of prior knowledge (Haidet et al., 2004) and active knowledge construction (Schmidt et al., 1989).

As described, average GPA and MCAT scores for the reporting classes were similar; therefore, it is likely that TBL contributed to the improved performances.

It is possible that TBL improved students' preparedness by encouraging them to keep up with the assignments rather than 'cramming' before exams. Furthermore, peer pressure to contribute to team discussions appeared to enhance performance.

There are a number of additional factors that might have contributed to the overall improved performances on the unit exams, NBME subject exam and licensing exam. These include: 1. The inclusion of clinical application exercises in learning topics for discussion during team encounters, 2. The improved quality of quiz and unit exam questions written in collaboration with clinical faculty, 3. Incorporation of high quality problem-solving and clinical reasoning questions obtained from various textbooks, Web sites, etc. into our team discussions, 4. Early clinical exposure through preceptorship, and, 5. Student's critical thinking and problem solving ability were likely improved.

Dinan (2002) reported that in undergraduate chemistry courses the use of modified TBL has resulted in significantly higher grades on standardized tests and fewer failures compared to the same course taught by traditional lecture-based means. One of the major benefits of TBL in that context was retention of academically weaker students.

We tabulated the anatomy course grades to evaluate the effect of TBL on student performance as a whole and how it might have helped the academically weaker students. Following the implementation of TBL curriculum, the students failing the course drastically diminished from 14 (2003) to 0-5 (after 2005). Furthermore, proportions of students receiving a pass (70%) decreased with a significant increase in number of students receiving high pass and honors (above 80%). What made the difference? As suggested by Dinan (2002), we believe that the combined use of reading assignments and quizzes served to elicit more students who studied anatomy on a daily basis. Furthermore, to insure individual accountability, individual exam grades are weighted more heavily than the group grades (85% to 15%). Using TBL, McInerney (2003) reported enhanced critical thinking and long-term retention in an undergraduate microbial physiology course. Our results also support the concept that providing an opportunity to learn in context with clinical cases, and discussing the cases as a team allowed deeper learning, better retention and improved performances.

CONCLUSION

Our attempt to incorporate modified TBL in medical human anatomy has been a positive and successful experience. We were able to cover course content and improve student performance despite reduction in course time allotted in the new curriculum. With continued refinement of "content specific learning topics" for team encounters and additional clinical application cases for inter-team discussion, we anticipate that might assist students to retain their knowledge of anatomy for application to their clinical years. We observed that the students were more engaged in their learning as the teams facilitated active learning as well as peer teaching.

Before choosing and instituting a formal TBL strategy, several circumstances should be evaluated. 1) whether or not it is the best approach for the subject material, 2) is there enough time available in the course for students to work both in class and out of class to evolve into a cohesive team, 3) is the task (course content) complex enough to justify team formation, and 4) will it help development of skills such as efficiency, critical thinking, higher-level reasoning, and teamwork.

There are other questions that require practical assessment. Are there administrative higher powers to "champion" the new approach? Are there teachers who will not do well with this strategy, but be successful lecturing? For a given course, are there other pedagogical approaches more appropriate than TBL? Will the team-based method negatively affect development of individual skills? Finally, does the TBL pedagogy during post secondary education enhance students' commitments to lifelong learning?

When the new curriculum was planned in 2004 with a reduction in curriculum time for anatomy, we did not see it as a problem but rather as an opportunity to implement TBL. To further our resolve, we were concurrently faced with a reduced number of teaching faculties. With the implementation of TBL in anatomy we addressed both the decreased course time and faculty. By piloting TBL in our first year, we learned which aspects of TBL were best suited for our students.

TBL gave us an opportunity to teach a subject that is rich in content to the satisfaction of students, faculty and the administration. TBL particularly excited us because of the empowerment it gave the students in their learning process and the associated opportunity it gave us to watch the students' blossom.

REFERENCES

Dinan F.J. 2004. An alternative to lecturing in the sciences. In: Michaelsen, LK, Knight, AB, Fink LD. (Eds.) *Team-based learning: a transformative use of small groups in college teaching*. P 97-104.Virginia, Stylus Publishing, LLC.

Haidet, P., Morgan, R.O., O'Malley, K.J., Moran, B.J., and Richards, B.F. (2004). A controlled trial of active versus passive learning strategies in a large group setting. *Advances in Health Sciences Education* 9: 15-27.

Herreid, C.F (1998). Why isn't cooperative learning used to teach science? *Bioscience,* 48, 553-559.

McInerney M.J. (2003). Team-Based learning enhances long-term retention and critical thinking in an undergraduate microbial physiology course. *Microbiology Education Journal,* 4, 3-12.

Michaelsen, L. K., Knight, A. B. and Fink, L.D. (2004) (Eds.). *Team-based learning: A transformative use of small groups in college teaching.* Sterling, Virginia, Stylus.

Michaelsen, L.K. and Sweet, M. (2008). Fundamental principles and practice of team-based learning. In Michaelsen, L.K., Parmelee, D.X., McMahon, K.K., and Levine, R.E. (Eds.) *Team-based learning for health professions education*, 9-34. Sterling, Virginia, Stylus.

Michaelsen, L.K., Parmelee, D.X., McMahon, K.K., and Levine, R.E. (2008) *Team-based learning for health professions education*, Sterling, Virginia, Stylus.

Nieder, G.L., Parmelle, D.X., Stolfi, A., and Hudes, P.D. (2005). Team-based learning in a medical gross anatomy and embryology course. *Clin. Anat.* 18, 56-63.

Schmidt, H. G., De Grave, W. S., De Volder, M. L., Moust, J. H. C., and Patel, V.L. (1989). Explanatory models in the processing of science text: The role of prior knowledge activation through small-group discussion. *Journal of Educational Psychology*, 81, 610–619.

Vasan, N. S. and DeFouw, D. (2005). A successful initial experience with modified Team Learning strategy in medical gross anatomy course. *Medical Education.* 39, 524.

Vasan, N. S. and DeFouw, D. and Holland, B. (2008). Modified use of Team-Based Learning for Effective Delivery of Medical Gross Anatomy and Embryology. *Anatomical Sciences Education*, 1, 3-9.

Vasan, N. S., DeFouw, D. and Compton, S. (2009). A survey of student perceptions of Team-Based Learning in Anatomy curriculum: Favorable views unrelated to grades. *Anatomical Sciences Education*, 2, 150-155.

Vasan, N. S., DeFouw, D. and Compton, S. (2011). Team Based Learning in Anatomy: An Efficient, Effective and Economical Strategy. *Anatomical Sciences Education,* 4, 333-339.

Vasan, N.S, and DeFouw D. (2007). The use of reading assignments and learning issues as an alternative to anatomy lectures in Team-Based Learning curriculum. In: Michaelsen, L.K, Parmelee, D.X, McMahon, K.K. and Levine, R.E. (Eds.) *Team-Based Learning for Health Professions Education*, P 169-175. Stylus Publishing, Herndon, Virginia, USA.

Vasan, N.S. and DeFouw, D. (2012). Team-Based Learning: Pedagogy for the 21st Century-Effective, Efficient and Economical. In: "*Pedagogy: New Development in the Learning Sciences*" P 271-295. (Robyn Gillies, Ed), Nova Science Publishers, Hauppauge NY.

In: Education in Anatomical Sciences
Editor: Paul Ganguly

ISBN: 978-1-62618-488-6

Chapter 5

NEW HORIZONS IN MEDICAL EDUCATION TECHNOLOGY

Volodymyr Mavrych * ***and Olena Bolgova***
Department of Anatomy, St. Matthew's University, School of Medicine, Grand Cayman

ABSTRACT

Within the last few decades there has been a paradigm shift in the academic delivery system for a medical education. Blackboards and renown textbook classics are being phased out in favor of PowerPoint presentations and subject based e-books. Tablets and laptops are now mandatory, and it is virtually required to have an online student schedule and electronic based curriculum. With a lesser emphasis being placed on cadaver dissection, there has been an emergance of plastinated specimens' supplimented with virtual 3D modules. With an increasing number of admissions, the larger medical universities prefer multiple choice exams as the gold standard evaluation, making oral and written examination more and more obsolete.

The lecture hall didatic method is still prevelent today, although with the increasing popularity of the PBL approach, more and more incoming students are required to be computer savory. Courses promoting the value of evidence based medicine are adamant about self-guided research, whether this study is for academic enrichment or the possible potential for publication. As the temper of the times shifts from the classical model of pen and paper to the newly favored system of computer interface, so too must today's medical graduates adapt to the ever constant technological advances in medicine.

1. COURSE DELIVERY SYSTEM

The educational core in the modern medical school is based upon an online Course Management System (CMS), also known as a Learning Management System (LMS) or a Virtual Learning Environment (VLE). Many universities use it as their platform to conduct

* E-mail address: prof.mavrych@gmail.com.

fully online courses, the popular systems being Angel (Blackboard Learn), Moodle and SimplyDigi course management systems.

Educators can then use these web applications to create an effective online learning site, one that handles and manages teacher-student communication as well as the delivery of coursework and online examinations [9]. The typical e-learning web portal consists of the home page with a complimentry course catalog, events calendar, list of global announcements, and the accompanying library resources.

Professors can then publish courses and place them in an easy to use online catalog. Students subsequently log into the CMS with their browser, select the desired course from the course catalog and simply complete enrollment. The CMS in turn will track the user activity, **monitoring the learner's progress, and the level of compliance** with the completion of all course assignments.

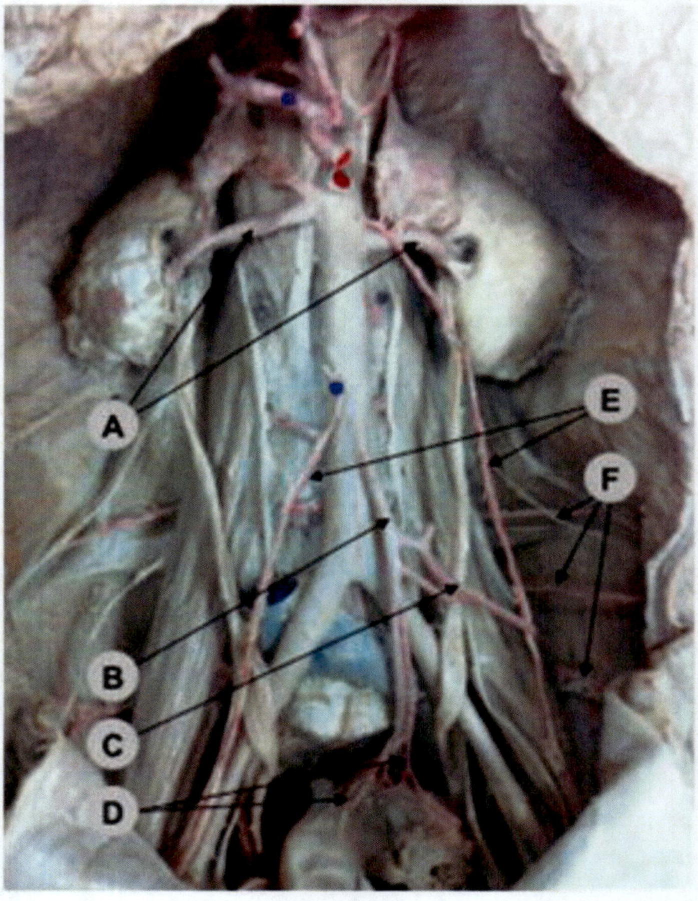

Figure 1. An example of the lab image posted online in Developmental and Gross Anatomy Course.

One increasingly popular option is the ability to embed real time lecture broadcasts into the relevent course website, allowing students to view the recorded lectures in their own free leisure time. This one invaluable feature is essential to the students who are absent during a given class, or the diligent learners that need additional time and reinforcement for that given lesson. Among the more popular programs commercially available are Panopto, Tegrity, EchoSystem, and Accordent Classroom Recording. These timestamp technologies are not only furthering the way educators provide lectures to their students, but they also allows professors to fascilitate on-demand guest speaker presentations and interactive symposiums over very large distances [18].

A well-organized subject provides students online with the main course information, comprehensible syllabus, various teacher presentations, optional review questions, and any recommended textbooks with online resources. Essential to any anatomy course is the accompanying lab component, and many medical schools give students access to a database of lab images, online interactive dissectors, prerecorded videos, as well as an extensive question bank upon the available specimens (Figure 1).

The CMS provides online reports for every learner and keeps track of their performance across several types of training activities. In addition these interactive programs serve as a model of effective communication between professors and students via constant activity on the online forums and through the exchange of email. It is recommended to equip each lecture hall with projectors, flat screen TVs or smart boards, as they all provide great opportunities to make lectures interesting and inspirational.

2. Resource Sessions

Formal lectures: Despite an archaic and outdated discussion about effectiveness, a formal lecture is still the backbone of many anatomy courses. Similar to the way that many instructors show a certain prowess with teaching the formal lecture style, so too do students gain a particular comfortability in learning under this format. As the professor gauges the class comprehension as a whole, the students benefit from the pressure of feeling responsible for knowing all the answers. Regardless of alternative preferences and multiple criticisms, the fact that the formal lecture persists today is a testament to its value as a current teaching necessity [6].

Formal lectures are vital to teaching the most quantitative and systematic sciences, especially for the intensive professional training courses such as medicine, engineering and law. Such rigourous programs impart a myriad of textual science, delivering class material in an organized fashion, equipped to address a large student audience. On the other hand, as in many cases, formal lectures provide limited interaction between instructors and students limited to the disadvantages of one-way communication. Passivity during lecture time is counterproductive; it can bore students into apathy and disinterest, with the tendency to scribble down lecture material without reflection or questioning. Such sentiments are perpetuated in the larger class sizes, furthered manifested by the pervasive feelings of anonymity and isolation.

This leads to the simple questioning on how to make a lecture more effective. A quinissential problem revolves around maintaining and captivating the student interest, especially given longer time periods of sitting and listening. The current teaching technologies are designed to address this problem, stimulating student interest by the use of drawing technology, video fragments, and "moving" images in smart boards. Using various clinical scenarios and question based discussion during lecture time; students are placed in the active role when given questions to tackle via a remote control responder. Clicker technology integrates a gaming approach that encourages students to participate anonymously. The benefits of active learning are widely acclaimed in higher education. According to Guthrie and Carlin, modern students are primarily active learners, and lecture courses may be increasingly out of touch with how students engage their world [13]. Chickering and Erhmann, early proponents of active learning, designated "encourage active learning" as one of seven principles of good practice in higher education [7]. Formal lectures are extremely well-suited to the transmission of conceptual and systematic knowledge, but of course, lectures are co-dependent on the student individual effort. There is a certain desirability today in the realm of healthcare for physicians who acquire a higher level of thinking through the apprenticeship of self-directed studies.

TBL sessions: Team based learning (TBL) is an institutional group approach towards problem solving, a learning methology which is implemented by multiple professional schools. Academic stress is placed on a communal learning style, where the primary learning objective requires constant group communication. Working as a cohesive unit each individual team uses previously learned concepts to tackle unfamiliar medical scenarios, with the benefit of immediate feedback towards all possible solutions. Larry Michaelsen, the central figure in the development of the TBL method, popularized the educational strategy that he developed for use in academic settings at University of Oklahoma. TBL provides the effectiveness of small-groups working independently in classes with high student-faculty ratios (up to 200:1) without losing the benefits of faculty-led small groups with lower ratios (6-7:1).

The role of an instructor in the TBL learning process is crucial. It's his responsibility to set learning objectives, select content and address learners' questions [15]. For maximal effiency, the instructor should distribute students by sections prior to class, avoiding membership coalitions which are likely to interfere with the development of the group cohesiveness. As the students work together periodically, groups evolve into a self-managed and truly effective learning team. For this reason, students should stay in the same group for the entire course. Each group should ideally contain a mix of demographic characteristics, varying the assortment of gender, age, and ethnicity. It is prudent for the TBL instructor to be well organized and prepared, as the benefits of group based learning are heavily reliant on the distribution of resource material and learning aids. In this regard the professor's success is based on his or her ability to identify relevent casefiles and anticipate how a student's will apply the key concepts. Classically, TLB consists of a repetitious sequence of 3 phases [14]. In the first phase, students independently study outside of class to master the relevent examination material and provided objectives. Classwork is stressed during the second phase, where individual students complete multiple-choice exams to assess their readiness with the subject knowledge. Groups of 6-7 learners then re-take this exam and turn in their consensus answers for immediate scoring and posting. The IF-AT answer cards is a convenient way to provide students with the feedback (Figure 2): they scratch off the covering of one or more boxes in search of apteryx sign, a mark indicating the correct answer.

If they find the mark on the first try, they receive full credit. If not, they continue scratching until they find the mark, but their score is reduced with each unsuccessful scratch. This allows teams to receive partial credit for applying proximate knowledge. Learner motivation is best observed when the individual student opts to use the TBL session as an assurance approximation of there grasp on the unti fundamentals. In the final phase, which spans several class periods, collaboartion is needed between groups to complete in-class assignments that make use of previous phases experience, as well as the identification of learning deficiencies. At designated time intervals, all groups simultaneously share their answers with the entire class for an easy comparison measure with immediate feedback. This stimulates an energetic total-class discussion with individual groups defending their answers to each other. Here the teacher's role shifts from being a source of primary exposure, to that of reinforcement in the consolidation of learning.

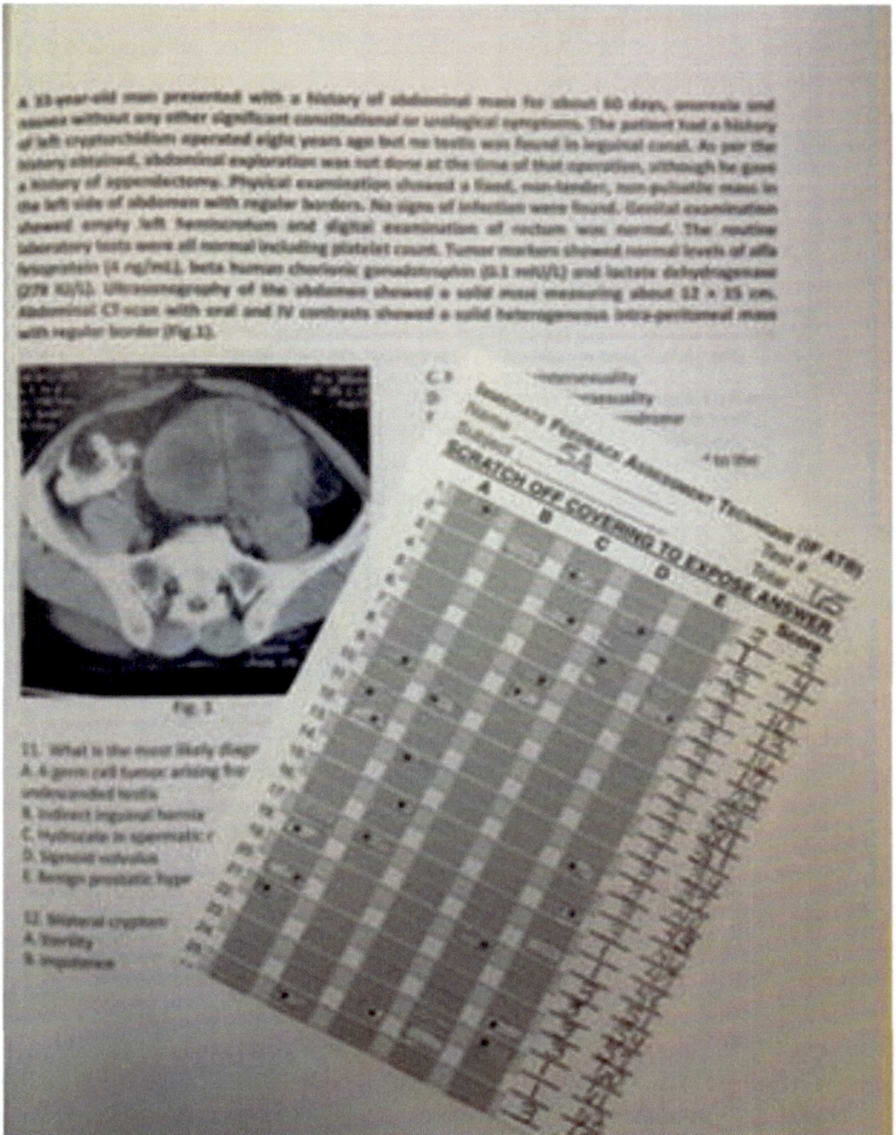

Figure 2. A page of a TBL test booklet with IF-AT answer card.

PBL sessions: Problem based learning is used in many medical schools in the UK, US, Canada and worldwide. The idea behind problem-based learning is that students acquire experience in addressing real-world problems within smaller groups. Fascilitators for each section will progressively disclose case based information as their team mates identify, analyze, and interpret the important case facts. The tutor is the responsible party in PBL, whose goal is strive the fundamentals and clarify misconceptions needed to progress forward with each case situation. Instructors are needed to construct scenarios in such a way that they will stimulate learners to pursue relevant learning issues and facilitate small group discussions giving feedback and guidance as necessary.

An Integrated curriculum is a much newer approach to medicine. As opposed to the traditional medical curriculum, which separates subjects such as embryology, physiology, pathology and anatomy, the integrated curricula combines multiple disciplines over the course of the first two years (Jonas 1989). The main idea here is to profess a systems based approach. When students are initially exposed to a new bodily system, such as Cardiovascular, they will be exposed not just to its anatomy, but also to its physiology, biochemistry, pharmacology, and pathology simultaneously. This new integrated approach encourages early patient contact and self-directed learning. Much of the teaching basis is grounded in lectures, supplemented with tutorials and students discrecion with self-directed work. This curriculum influences participants' motivation, learning and thinking, and makes a shift from a teacher-centred orientation to a student-centred approach.

3. Gross Anatomy Laboratory

The Gross Anatomy lab is the hallmark building of the medical school guided tour, yet it's more importantly serves the working place for the 1st year medical student. Since antic, anatomy has provided a foundation for medicine, the medical school basement where student are exposed to tissue specimens and cadavers.

Dissection: Most of medical schools incorporate dissection and stock a fully equipped anatomy lab. It typically incorporates the following structural components: the main anatomy lab itself with 10-25 worktables designed for 6-8 students at each table with adjacent morgue, male and female locker rooms, faculty office, virtual Gross Anatomy computer lab, general storage room and AV storage room. The mechanical, electrical and plumbing systems should be sophisticated, in keeping with the demanding nature of Gross Anatomy lab usage. The overall finishes of the lab space should be designed for maximum durability. Ideally each dissection station need to be equipped a networked computer terminal, white board, lockable storage drawer for instruments/gloves hand, washing sink and dual exam lights. Wall mounted 55” - 65” display monitors are typically positioned between student dissection stations and are used for instructor based presentations and demonstrations.

Controlling formaldehyde emissions in a Gross Lab is the main problem of any anatomical workstation. Anatomical dissection, especially in the long-term semesters with 7-11 mounts, is a well-recognized source of toxicity. Formaldehyde is classified as a group 2A human carcinogen (IARC classification). Published workplace monitoring results suggest that formaldehyde exposures during anatomical dissection can approach and periodically exceed OSHA’s PEEL and STEL levels [10].

Plastinated specimens: Younger medical schools typically prefer plastinated specimens over cadavers in their Gross Anatomy labs. Plastination is an amazing technique of human body fixation that replaces an organ’s body water with acetone and plastic, and it is relatively inexpensive in comparison to using human corpses. Like most inventions, plastination is simple in theory: decomposition must be halted in order to make a specimen permanent. Cell enzymes released after death trigger an initial decomposition. This process of abrupt decay is problematic as it undergoes putrefaction leading to a full bacterial colonization. Plastination process deprives the bacteria by removing nutrients from the tissue and replacing the water and fat with polymers. However, bodily fluids cannot be replaced directly with polymers, because the two are chemically incompatible.

Gunther von Hagens is the innovator and father of plastination: initially water and fatty tissues are replaced with acetone, a solvent that readily evaporates. In the next step the acetone is replaced with a polymer solution, and the specimen is placed in a vacuum chamber. Next, the pressure is reduced to the point where the solvent boils, allowing the acetone to be suctioned out of the tissue at the moment it vaporizes. The resulting vacuum within the specimen causes the polymer solution to permeate through the tissue. This exchange process is allowed to continue until all of the tissue has been completely saturated. The duration of this process can take weeks for whole bodies, but it's a matter of only a few days for thin slices. Because of such a technology, plastinated specimens are very expensive but their quality is just amazing (Figure3).

Figure 3. Plastinated human body specimens, EB 2012 San Diego.

Since plastinated specimens don't require any special preservation techniques they can be used indefinitely. Gross anatomy lab consists of the main anatomy lab itself with 6-10 workstations designed for 6-8 students, faculty office, virtual Gross Anatomy computer lab, and storage rooms. It is custom for the Professor to provide a short instruction at the beginning of the lab session followed by subsequent student rotations among the various stations. For the instruction, the wall mounting video camera with the 10x or more zoom and hands-free wireless microphone should be used.

The lab also requires a few projectors or wall mounted 55" - 65" display monitors. Every station has a set of professionally dissected plastinated specimens with detailed directions. The instructions contain a list of anatomical structures, which students should be able to identify and review. Since every specimen has some specific features, it is reasonable to make student teams rotate, and using plastinated specimens prevents medical students from hyper exposure to formaldehyde.

Virtual dissection table: Anatomage Table is a computerized LCD life-size virtual dissection table. The table was originally designed for anatomy courses without cadavers, but may be complement for cadaver based dissection courses. It provides a realistic visualization of the 3D anatomical details of a fully dissected individual. A full body gross anatomy model was rendered from CT scan data. The Table can also open any data from CT, MRI, and ultrasound scanners [3].

Additionally 2D photographic images, or even full presentations, can also be displayed on the Table. Anatomage allows students to visualize skeletal tissues, muscles, organs and soft tissue. Students can use the program to cut the body, peel off soft tissue or fully remove an organ with their fingers. And unlike cadaver dissection, the students can undo and redo the simulated dissection as many times as necessary. Virtually slicing, layering, and segmenting different cross-sections allows for a user based instruction which is both natural and intuitive. The full size of the body table allows easy integration into small group learning programs. The table allows student teams to replicate a dissection experience, as they stand beside and interact with the volume, turn patients, remove portions or completely change the view with the swipe of a finger.

The life size 3D rendering displayed on the Table allows for case reviews with a complete medical team. A full team of specialists may review any clinical case together, discussing different images and the planning involved for a given surgical procedure. The Table also can be used for visual communication with patients, allowing explanation for a particular procedure or consultation. Demonstration for the 3D patient's own scan in life-size is a very impressive and effective method of presentation.

4. 21st Century Modality for Living Anatomy Program

Clinical anatomy represents the trend to integrate the basic sciences with its hospital applications. This approach is particularly important in the current medical curriculum, where details of the anatomical subject shy away from a more dence clinical application [11]. But this trend is somewhat unfortunate, given that medical students should be prepared to enter the clinical environment from the very start of the medical education.

Surface anatomy forms the obvious connection between the gross anatomy and the broader clinical practice, forming the bridge for physical examination. Knowledge of the surface anatomy of the thorax and abdomen is particularly important because these two regions are frequently tested from its subject discussion to its procedural application in auscultation, percussion and palpation. In particular, the palpation of the main bony key landmarks is useful in regards to different type of punctures, injections, and local anesthesia technics. Along with the regular studding of surface anatomy, the projection technique and body painting are both commonly implemented. Both of them use the body surface as the imaging medium for presenting internal anatomy. These technics use either commercially available images or hand painted displays to enhance students' learning of surface anatomy. Radiological and endoscopic anatomy are both parts of clinically orientated anatomy, have been adopted by many medical schools because of the increasing reliance on medical imaging for diagnosis [12]. They provide tools for students helping them achieve a real understanding of the application of anatomy in the clinical context. It's practically important to arrange specific station in the gross anatomy lab for radiographic films, CTs, MRIs, ultrasound and endoscopic images.

Plain radiographs are common to both local practice and hospitals, making x-ray interpretation one of the most basic skills for every physician. Plain radiographs are extremely useful for verifying traumatic accidents and in identifying bony fractures and joint dislocation. Learning of this type of images should be emphasized in Back, Chest, Head and Neck, as well as material for the Upper and Lower limbs. Abdominal material should be highlighted with gastrointestinal contrast to demonstrate esophagus, stomach, small bowel, and large bowel. A urological contrast study is the good way to demonstrate the pelvicoureteric system, bladder and urethra. Angiograms, which demonstrate coronary arteries, branches of the aortic arch, and the abdominal aorta, as well as, branches of the carotid arteries and venogram of dural sinuses may be used in the proper parts of the course. Mammography is a routine examination and should be demonstrated along with the female breast specimens.

Computed tomography (CT) and *Magnetic resonance imaging* (MRI), are two of the utmost important radiological techniques for visualizing the internal human landscape. Contrary to their differences in both indication and procedure, both the CT and MRI share many similar features. The best way to understand the location of the different structures using these images is by learning the multitude of axial body sections. The plastination technique allows receiving and saving body sections in any direction, but normally there are 3 classical anatomical planes used. Students start working with the axial section of the body and continue with CT and MRI. CT and MRI images are the most intensively used in the Thoracic and Abdominal material, but the MRI of the shoulder region and knee joint should be presented as well, since they provide basement for a diagnostic of tearing of the supraspinatus muscle tendon (or subacromial bursitis) and damage of the different intra articular structures in the knee (mostly cruciate ligaments). Students may be presented with 3D MRI reconstruction, which is used now for planning a surgery (e.g. brain, pelvis or heart) and reconstructing computerized versions of an anatomical atlas. Due to time constraints, a normal curriculumn does not allow using specialized types of MRI in the Gross Anatomy course. With that said, so such images, like magnetic resonance angiography, Magnetic resonance gated intracranial CSF dynamics (MR-GILD) and functional MRI should be presented in later semesters such as in a neuroscience or separate imaging course.

Ultrasonography is a popular tool widely used for all medical fields today, and ultrasound images should be present in Gross Anatomy lab along with radiographs, CTs and MRIs. Since ultrasound is a non-invasive method, a machine may be localized to a student station to provide classmates with necessary experience in handling and creating an ultrasonogram. Developments in ultrasound technology, including various probes size and frequency range, permit different anatomical areas to be scanned. Traditionally ultrasound is used for an Abdomen examination: where students can easily identify kidneys, gallbladder, abdominal aorta and IVC. In the Neck material, a thyroid gland and sublingual lymphatic nodes are primary targets for identification. Doppler ultrasound images may be also demonstrated to identify sites of blockage in blood vessels.

Endoscopy (gastroduodenoscopy, colonoscopy, bronchoscopy) is a very useful nonsurgical procedure used to examine cavities of empty organs. Endoscopic images of the esophagus, stomach, duodenum, large intestine and bifurcation of the trachea may be used on the radiographic station to reinforce students understanding of different clinical applications, which are discussed during the lab session.

5. Professional Skills

Standardized patients: A simulated patient or standardized patient (SP) is an individual who has been trained to consistently and accurately simulate an actual medical condition or disease, including a patient's history, physical findings, personality and behavior. Standardized patients have been successfully used in medical education, nurse training, evaluation, and research. SPs are trained to evaluate learners' performance, provide feedback on professionalism, communication and clinical skills, which may include physical examination techniques [5]. Thus, an SP can serve as a patient, teacher and evaluator and allow learners the opportunity to practice repeatedly in a safe, and controlled realistic atmosphere. SPs assessment has now been incorporated into U.S. national licensure exams (USMLE Step 2 Clinical Skills) for medical students [19].

Educational activities involving SPs can be formatted in multiple fashions, such as one-on-one sessions, communication skills practice, or small group and large group sessions. The use of SPs in the medical school has several advantages: it allows students to learn clinical cases at scheduled time and make possible direct comparison of the students' clinical skills [8]. The supervision of medical students by SPs reduces the need for clinical faculty and shows the way, how students need to manage situations alone in a real clinical setting.

Medical manikins: Medical manikins are regularly used in the training of medical students. Medical manikins range from simple plastic, and have digitilized in their complexity to now incorporate computers, printers, and lights. These units ensure that students are up to speed on crucial medical procedures. There are a few standardized models to the medical manikin. One common example is that the full size and torsos of an infant, child, and adult are the most realistic simulation dummy and they are very effective for training in CPR and choking techniques. In a more clinical setting, basic patient care manikins are ideal for showing medical students the correct ways to move a patient or even how to perform certain procedures. Advanced intramuscular injection training model may be used to train students in injection protocol. Such models contain the main structures of the gluteal region: bony key landmarks, gluteal muscles, and sciatic nerve. The gluteus maximus muscle can be taken away given possibility to observe internal structure and make sure about position of sciatic nerve. An obstetrical manikin is a pelvic model with a full term newborn and placenta. It provides students with realistic practice of the emergency childbirth procedure. It comes with replaceable spare vulva, disposable umbilical cords, clamps, powder to make simulated blood, a modular pregnant belly overlay. A newborn size, a head with fontanels and cranial sutures provide students with good realistic experience.

Using the model, students can find the most important pelvic landmarks and practice the Leopold's maneuver. The arrhythmia simulator manikins have a very special function for those first aid workers and medical professionals who are likely to come across this type of situation.

The blood pressure training arm is a lifelike, adult arm with an electronic trainer designed for training the blood pressure measurement. The specific features of this manikin are palpable antecubital pulse; blood pressure trainer with LCD guided operation, which allows checking both systolic and diastolic pressures (systolic, diastolic, heart rate and auscultatory gap are programmable).

First aid arm manikins both look and feel like a real arm. A synthetic blood flow can be established from either the cut on the forearm or from the severed thumb, allowing for either the venous or arterial bleeding rates to be regulated.

Multi-venous IV training arm is the lifelike reproduction of a complete brachial venous system. It may be used for training of peripheral intravenous therapy skills and procedures including IV cannulation, infusion, IV bolus, IV push, site care and maintenance. Catheterization/Enema simulator male/female may be used for basic trainings in urethral catheterization, perineal care and enema. Students receive a realistic feeling of catheter insertion because of the soft and anatomically correct genital organs. Also, the organ unit can be used independently without the lumbar torso, allowing training with SPs or full body manikins in a clinical environment.

Heart and lung sounds manikin allows teaching of auscultation on the adult torso with palpable anatomic landmarks. Appropriate sounds are heard when the student places the virtual stethoscope at corresponding regions of auscultation for the heart. Furthermore, students can listen to the appropriate lung sounds as the bell of the stethoscope is moved across the front and back of the chestwall.

Intelligent digital network-based ECG simulation teaching system is the latest ECG simulation manikin used for teaching, practice and assessment. This style of manikin relies on innovative technology to further develop student competency with cardiograms, while opting for a more interesting teaching medium.

6. Computer Assisted Education

Today's medical classes almost all unanimously have virtual online components, and there are a plethora of options for both teaching and studding an anatomy course. Whereas the classical approach to learning once relied heavily on textbooks, atlases and learning reviews, our current web technology allows there to be an exciting transformation. It can be problematic for an anatomy leaner to try to form a three dimensional mental picture by looking at images in books, screen slides, or static images on a monitor. The optimum 3D learning experience prevails when dissecting the human body itself or through plastination, but not every student has access to these resources. For those who must learn or relearn human anatomy without access to cadavers, there's a critical need for a learning tool that presents human three dimensional body images. There are two different approaches in the presentation of three-dimensional images: stereoscopic vision and rotation. Stereoscopic vision we use for our everyday depth perception and it's the basis for all 3D images and movies. These visual aids all require special equipment (3D glasses and screens). There is another method via the rotation of the object. Object rotation gives our seeing brain a succession of images that it synthesizes to create a three dimensional mental image. Once we have that three-dimensional image, any two-dimensional image of the object that we see later is superimposed into the three-dimensional domain.

Visible Body 3D Human Anatomy Atlas: Visible Body 3D Human Anatomy Atlas serves as multiplatform software available for PC, Mac, Android, iPad and iPhone. Visible Body 3D Human Anatomy Atlas includes complete, fully interactive, 3D human anatomy models with easy-to-use controls and more then 2,500 anatomical structures [20]. It contains all the major organs of the male and female body, with hundreds of thumbnails that launch preset views of anatomical regions and body systems. With Visible Body 3D Human Anatomy Atlas, students can move the model in three-dimensional space, hide, rotate, see through, and explore human anatomy search for and locate anatomical structures by name. Also there are few detail 3D models are available: muscles, heart and circulatory, brain and nerves system, digestive, respiratory reproductive and urinary systems.

Acland's DVD/Video Atlas of Human Anatomy: Acland's Video Atlas uses fresh, unembalmed specimens that retain the color, texture, mobility and beauty of the living human body. A concise synchronized narration runs throughout the video. As each new structure is shown, its name appears on the screen. Several unique features make this video atlas different from other available anatomical teaching tools: exquisite dissections and fresh human specimens in their natural colors, beautiful video with clear narration, on-screen titles [2]. There is a self-testing feature at the end of each section. Acland's Video Atlas is a timesaving aid to first-time learners, an effective way to relearn anatomy for clinical rotations, and a time-efficient tool for review at test time. For clinicians in training and in practice it assures a swift renewal of anatomic knowledge. For instructors Acland's Video Atlas shortens the time needed to provide satisfying explanations of three-dimensional structure.

A.D.A.M Interactive Anatomy: A.D.A.M Interactive Anatomy is an online recourse (originally it was CD version with online key), which includes detailed graphics combined with A.D.A.M.'s Multimedia Encyclopedia with over 3,600 articles covering conditions, disease states, procedures, treatments, surgeries, tests and more, over 3,000 illustrations, diagrams and photos and over 120 multimedia videos [1]. The 3D Library contains 8 fully interactive 3D models, including models of the skull, brain, ear, eye, and heart, lungs, male and female reproductive systems. A.D.A.M Interactive Anatomy beyond anatomy courses includes a lot of clinical applications and concepts for your allied health, nursing, continuing medical education (CME) and medical assistance courses. Clinical animations and viewer screens contains 28 professionally produced animations cover topics related to physiology, disease, and surgery.

Primary Pictures: This web developer provides a wide variety of products available via online subscription or DVDs. Primary Pictures provides wide variety of 3D anatomy CD-DVD based products: Anatomy for Acupuncture, Anatomy for Chiropractic Spine, Primal Interactive Human, Head and Neck Anatomy for Dentistry, Anatomy for Dental Hygiene, 18 Regional Edition DVD-ROMs, Interactive Functional Anatomy, Anatomy for Otolaryngology and Head and Neck Surgery, Anatomy for Speech Language Pathology, Head and Neck Anatomy with Special Senses and Basic Neuroanatomy, Anatomy for Massage and Manual Therapies, MSK Ultrasound, Anatomy for Urology, Radiological Cross-Sectional Anatomy with Multidetector CT, Primal Interactive Human, Anatomy for

Yoga [17]. Online Anatomy and Physiology course consists of 19 fully interactive modules, which covering the entire content of the classical anatomy and physiology college course. Each module has realistic case studies, giving real medical context and relevance, as well as a dedicated section on aging. Anatomy and physiology online course engage using of different types of media such as sound/voiceover, interactive 3D images, animations, labeling and layering, pronunciation guide. There are interactive quizzes at the end of each section with different types of questions.

Thieme Dissector: Thieme Dissector is a web-based platform for creating customizable dissection guides. Both the content and the structure of the dissection guide can be customized, allowing for a created dissection guide geared especially for specific courses and student groups [16]. Text can be easily edited and adapted to include an instructor's personalized step-by-step directions. Stunning, full-color images are from the Thieme Atlas of Anatomy. Additional 'pop-up' content provides additional information on muscle attachments and actions. Final, customized output is available to students online, on iPads, and on their local computer. Customization is a major key feature of this program, as it can be used on the default setting, or even fully adapted to the instructor's user preferences. With just a few clicks a complete Dissector can be accessible for online viewing or for download or printing by students.

AnatomyOne: AnatomyOne is an online course for supplementing human anatomy that contains illustrations with annotations and labels that turn on and off, prosection photographs, radiographic and histology images; over 1,000 USMLE-Style multiple choice questions and dissection guide labs [4]. Comprehensive text with supported hyperlinks serves as a regional, systemic, and clinical anatomy comprehensive catalog. Instructors may create a course syllabus that links to AnatomyOne resources (such as modules, customized atlases, dissection labs, customized glossary lists, and more) and embed it to the school Learning Management System.

iPad applications: A lot of iOs applications are available on the current market, lets identify few the most sufficient for the medical students: textbooks and application with real specimen photos, X-ray images: Netter's Anatomy Atlas, Grays Anatomy Student Edition, Surgical Anatomy, GAlab Thorax, GALab Pelvis, GALab Back and Neck, GALab Head, GALab Abdomen, UpperLimb, Arm, Radilogical Anatomy, Green Radiology. Quizez: GA Cases, GA_X-rays, CS Quiz, GAQuiz Head. Reviews: 100 GA_Concepts. 3D Application: Visible body 3D Human Anatomy, Muscle and Bony Anatomy 3D, Skeletal Anatomy 3D.

iTunes U: iTunes U app is a new application from Apple to provide a learning material for students by creating courses for iPad. Now it's even easier than ever to unleash the full potential of iPad in a classroom. iTunes U courses include all the items you would use in a traditional course such as a syllabus, handouts, presentations quizzes, documents, PDFs, ePub books, iBooks textbooks for iPad, audio and video, Web links and the best is about, everything works together. iTunes allows to add materials from free public resources, among them: audio and video content from museums, universities, cultural institutions (more then 500,000 in the iTunes U catalog). From the iTunes U app, students can casually listen to prerecorded lectures and synchronize note taking. Besides having an online activity checklist, the student can tell if an assignment includes watching part of a video. All it requires is one mouse click to go straight to the specific media segment. Instructors may send a message or create a new assignment, and students receive a push notification with a subsequent invoice sent to their university email.

8. Online Assessments

Assessment is the keystone of any teaching course. It's an evaluation of students' knowledge, and currently most of the quizzes and exams in medical school are computer based and integrated in the Learning Management System. Starting from weekly quizzes and finishing with NBME subject shelf exams, all assessments may be done on students' laptops or school desktops which serve as working stations. Respondus LockDown Browser is specific software for PC and Mac that integrates LMS (Moodle, Angel, Blackboard) with a lock down testing environment. Using Respondus LockDown Browser during the test, students are not allowed to print, copy, venture to another URL, or access other applications until the test is submitted for grading. NBME has transitioned to paperless and moved all the Basic and Clinical Science subject examinations to a web format, and they too now provide their own browser for online exams. Online assessments have a lot of the advantages, that not only save time, but save financial costs on buying ink and paper. LMS based assessments support all types of questions: fill-in-the blank, true/false, multiple choice, multiple select, matching, ordering, short answer, and essay. Questions may include graphics and media files using online HTML Editor, with the current format favoring multiple-choice. LMS specifies which students have permission to take the assessment, when the quizzes will appear in the course folder and the dates they will be available to students. An instructor can provide a password to an assessment with certain specifations given to a particular testing site. The instructor in turn can determine how students interact with the assessment: such as the number of the questions on the screen, how the questions are randomized, the ability to set a time limit, and even the capacity for submissions to be anonymous. LMS can give direct feedback after one question, a subset of questions or even the entire assessment bank. Here the learner is informed of both correct and incorrect answers, their achieved overall score, and any possible grader remarks.

During an LMS quiz, the proctor views the real-time assessment completion status. Individual student scores, high, low, mean and median scores are available for the instructor right after the submission. LMS presents a finalized summary of assessment question types, overall numbers and points assigned, and of course the assessment settings. An item analysis is available immediately after the assessment, and typically includes the Mean, Standard deviation, Standard error, Reliability, and Discrimination, Difficulty distribution. Lastly, an instructor can check possible technical mistakes before releasing students' grades in the LMS grade book. LMS also provides survey capabilities which can be used prior and post completion of the course. They allow an instructor to evaluate the base subject knowledge of a class, in addition to reviewing his or her teaching effectiveness. As a post-course tool, the surveys feature allows its administrator to observe the general student perspective regarding the course as a whole, with individual commentary on the instructor's performance.

References

[1] A.D.A.M Interactive Anatomy http://www.adameducation.com/aiaonline.aspx

[2] Acland's Acland, Robert D. FRCS. DVD/Video Atlas of Human Anatomy. Publisher: Lippincott Williams and Wilkins.

[3] Anatomage Table - *Virtual Dissection Table* http://www.anatomage.com/
[4] AnatomyOne http://www.anatomyone.com/
[5] Barrows HS, Williams RG, Moy, RH. 1987. A comprehensive performance-based assessment of fourth-year students' clinical skills. *Journal of Medical Education* 62: 805-809.
[6] Charlton BG. Lectures are an effective teaching method because they exploit human evolved 'human nature' to improve learning - Editorial. *Medical Hypotheses* 2006; 67: 1261-5. http://www.hedweb.com/bgcharlton/ed-lect.html.
[7] Chickering A. W., Erhmann S. C. (1996). Implementing the seven principles: Technology as lever. AAHE Bulletin, 49(2), 3 – 6.
[8] Cleland JA, Abe K, Rethans JJ. 2009. The use of simulated patients in medical education: AMEE Guide No 42. *Med. Teach.* 31(6):477-86.
[9] Ellis R., Calvo R.A. (2007), "Minimum indicators to quality assure blended learning supported by learning management systems" (PDF), *Journal of Educational Technology and Society.*
[10] Fleisher J. 1987. Medical students' exposure to formaldehyde in gross anatomy laboratories. *New York State Journal of Medicine* 87:385-388.
[11] Ganguly P.K. 2010. Teaching and Learning of Anatomy in the 21st Century: Direction and the Strategies. *The Open Medical Education Journal* 3:5-10.
[12] Ganguly P.K., Chan L.K. 2008. Living anatomy in the 21st century: how far can we go? *South East Asian Journal of Medical Education* 2 (2): 52-57.
[13] Guthrie R. W., Carlin A. "Waking the Dead: Using Interactive Technology to Engage Passive Listeners in the Classroom," *Proceedings of the Tenth Americas Conference on Information Systems,* New York, August 2004.
[14] Michaelsen L.K., Parmelee D.X., McMahon K.K., Levine R.E: editors. Team-Based Learning for Health Professions Education: A Guide to Using Small Groups for Improving Learning. Foreword by Diane M. Billings. Sterling [VA]: *Stylus Publishing, LLC*; 2007.
[15] Nieder GL, Parmelee DX, Stolfi A, Hudes PD. 2005. Team-based learning in a medical gross anatomy and embryology course. *Clin. Anat.* 18:56–63.
[16] Online Thieme Dissector http://dissector.thieme.com/
[17] Primary Pictures http://www.primalpictures.com/
[18] Sohn T. "Panopto Focus 4.0 Gets Moodle 2.0 Integration". *THE Journal.* Retrieved March 5, 2012.
[19] Van Zanten M, Boulet JR, McKinley DW, DeChamplain A, Jobe A. 2007. Assessing the communication and interpersonal skills of graduates of international medical schools as part of the United States Medical Licensing Exam (USMLE) Step 2 Clinical Skills (CS) Exam. *Academic Medicine* 82(10): S65-S68.
[20] Visible Body 3D Human Anatomy Atlas http://www.visiblebody.com/

In: Education in Anatomical Sciences
Editor: Paul Ganguly
ISBN: 978-1-62618-488-6

Chapter 6

Anatomical Education: Gross Anatomy (Part 1)

Stuart W. McDonald*
University of Glasgow, Glasgow, Scotland, UK

Summary

This chapter gives a comprehensive overview of the practical issues that require consideration by the staff running a course in gross anatomy. The general objectives of courses in gross anatomy are discussed along with the rationales that govern whether the practical elements of teaching should be delivered by dissection, prosection or some other means. The need to tailor the anatomical teaching to local requirements and resources is emphasized. The chapter also addresses many important logistical issues on the organization of practical anatomy classes, staffing and the presentation of the cadaveric material. Advice is provided on ways in which to enhance student learning and on how pitfalls may be avoided in running practical classes in gross anatomy. The article also considers the anatomy donors and how universities might mark their generous bequests. The author hopes that the ideas presented will be helpful to colleagues engaged in teaching gross anatomy.

Introduction: The Purpose of a Course in Gross Anatomy

For most students, the purpose of education in gross anatomy is to allow understanding of the spatial arrangement of body structures, how each part functions as an individual component and as part of a larger body system, thus contributing to the general functions of a living organism: respiration, feeding, digestion, movement, excretion, reproduction etc. For clinical groups there is a need to be able to relate surface morphology to underlying anatomical parts to inform diagnosis and treatment. Some groups might have a more specific

* Corresponding author: E-mail address: Stuart.McDonald@glasgow.ac.uk.

aim, for example, it might be possible to run an anatomy course for archaeological osteologists to allow them to better understand muscle attachments to bones and the nature of bony landmarks that might be used in bone identification. For this chapter, I was asked to describe how to run a course in gross anatomy. I hope that readers gain ideas from the reflections offered.

Constraints on Adoption of Ideas in the Literature

Although many papers have been written and presentations given on how to teach gross anatomy, and journals are even devoted to this topic, anatomists will generally find themselves unable to do more than adopt some of the ideas and practices that they describe, most probably in modified form. This is because education in anatomy in any given institution has to be tailored to suit the philosophy of the particular course being delivered, the amount of time available for anatomical teaching in the curriculum, the stage of the overall programme at which the anatomical teaching is to be presented, and the staff and facilities available to deliver it. I wonder how often anatomists feel frustrated at educational meetings when speakers expound upon good practice but the practitioners find themselves restricted by the circumstances at their own place of work? There is also a tendency for anatomists to write on educational curricula as if they were only teaching medical undergraduates or surgical postgraduates when in fact the more progressive institutions are delivering anatomy courses to a wide range of other students: dental, science and trainees in professions that support medicine such as physiotherapy, radiography, speech therapy, podiatry. The points made in this chapter will, I hope, be found to be applicable to any course in gross anatomy.

Dedicated or Mixed Courses?

Thought is sometimes given to organising a course to provide simultaneous tuition for more than one group of students. There are two draw-backs to such an approach. One is that the needs of the separate groups making up the common class differ. For example, consideration might be given to running a combined head and neck course for medical and dental students. The needs of medical and dental students with respect to knowledge of the anatomy of the head and neck, however, are somewhat different. The former require a greater emphasis on issues such head injury and otorhinolaryngology while the latter need a greater coverage of the three dimensional relationships of structures in the vicinity of the oral cavity, on the functional anatomy of these structures, and on the anatomical basis of local anaesthesia of the jaws and associated mucosae. When serving as an external examiner, I have encountered common courses for students of physiotherapy, of radiography and of health science, sometimes with all three groups being taught together. These courses worked reasonably well, but it is clear that an anatomical course for physiotherapists will require teaching on muscles and their movements that would not be essential for radiography and, contrariwise, the radiographers are likely to require a level of detail on osteology and on aspects of internal organs that may be imaged that would not be wholly appropriate for physiotherapists. The second drawback is the possibility of differing levels of ability between the groups of students. At the institution that was teaching health science students along with students of physiotherapy and of radiography, the entrance requirements for the last two groups was higher than for the first so that the health science students generally fared less well in terms of examination results. I was unable to gauge the effect of this on the morale of these students but it was likely to have been a dispiriting one, affecting both their professional confidence and their enjoyment of a university course. It must be very demotivating to be in a class where there is the clear evidence of school leaving certificates that the majority of the class members are more academically gifted than you. Attempting to form combined courses for two or more groups of students where professional needs are different or where general intellectual abilities are unequal is a sure recipe for dysfunction, at least to some extent. Where possible, it is better to run courses tailored to the needs and abilities of individual student groups.

The Choice of Teaching Method

Teaching by prosection was first suggested by Wakeley (1953). Until 1958 in the United Kingdom, the General Medical Council stipulated that all medical students must dissect the whole body (Sinclair, 1965). After that, other methods of teaching gross anatomy started to be introduced and their efficacy compared to dissection. In turn, of course, the loss of the requirement for dissection led to the erosion of the time available for anatomical teaching which has continued to such an extent that in recent years health professions internationally are expressing concern that anatomical training is unacceptably poor, at least in some quarters (Cottam, 1999; Turney, 2007; Gupta et al., 2008; Craig et al., 2010).

The decision on whether education on gross anatomy should be delivered by dissection, prosection, or by some other means, depends on the level of detail required by the students, on the availability of human material and on the time available in the curriculum. Clearly, a course studying general human biology does not need to dissect if all that is required is a grounding in the general layout of the human body. On the other hand, dissection of the head and neck and perhaps the thorax would not be inappropriate for dental students and dissection of all or part of the body could also be appropriate for medical students, as more detailed anatomical knowledge is required by these health professionals. In recent years, many computer-based learning packages have become available but my own view is that these are an adjunct for learning from cadaveric material rather than a replacement for it. Students need to see the real anatomy to be able to appreciate the scale, depth and variability of structures to more fully understand what they see on a computer screen.

A distinct advantage of dissection over prosection is that the former conveys knowledge that goes beyond presenting basic structure, which is valuable for certain student groups. For example, when dissecting the external carotid or the celiac arteries, students are surprised by the variability in the arrangements of the branches, and learn the important point to expect variations if carrying out surgery, as well gaining an idea of what these variations are likely to be. In head and neck, every time I run the dissection course, I realise something that I had not appreciated previously. For example, recently I was invited to give a lecture to a conference on the fascial planes of the head and neck. Shortly afterwards, in the dissection class, I realised that when I turned medial pterygoid aside to see the muscles attaching to the styloid process, this could be done readily by virtue of the loose connective tissue plane of the lateral pharyngeal space (Grodinsky and Holyoke, 1938).

Dissection, however, only works if a fairly strong regional or systems-based approach (see later) to the curriculum is adopted. For several years my own university adopted a problem-based medical programme which was not conducive to a dissection course, although some regional dissection was offered, mostly as special study modules. The majority of the practical classes were therefore based round the study of prosections.

Regional or Systematic Gross Anatomy

Dissection courses work best with a regional approach. Systems-based and integrated courses match better to a prosection-based programme, but a dissection-based course could be successful with careful handling. A systematic anatomical textbook is one in which the various body systems are described in separate chapters: skeletal system, joints, muscles, cardiovascular system, respiratory system, gastrointestinal system etc., etc. Division of the body into systems is a classification method, which like all classifications, is adopted to help us by aiding understanding. There is a certain mystique about a classification and inexperienced staff might not realise that in anatomy there is leeway in systematic classification whereby elements can be singled out or grouped together according to utility. For example, students could consider the "urinary system" and the "reproductive system" or, by combining them, the "genito-urinary system". The skeleton, joints and muscles could all be treated as separate systems or be brought together as the "musculoskeletal" or "locomotor" system. A dissection programme for the locomotor system, for example, would be possible if it were limited to the limbs. If the vertebral column is to be included, however, it is probably not going to be possible to expose the vertebral column and its relations until the thorax and abdomen have been dissected, and even then is unlikely to make a good class dissection. For the gastrointestinal system, dissection of the abdomen works well but demonstration of the pharynx and oesophagus needs to await dissection of the neck and of the thorax, if these regions have not already been exposed. Thus a series of primarily dissection-based practicals supplemented by prosected specimens could be appropriate. My feeling is that practicals with dissection and prosection mixed do not work well and my own solution would be to have dissection only for abdominal parts and to describe the non-accessible organs in a lecture. Separate courses for the cardiovascular and respiratory systems will not work well by dissection but a very satisfactory dissection of the thorax could be achieved by combining the two into one block of teaching, the cardiovascular and respiratory systems. Practicals to accompany a course that is strict about taking each system in turn would lend itself to a prosection-based approach, and probably even better a combined prosection and microscopy course. If dissection is preferred, really what needs to be done is to deliver the gross anatomy course by the most regional approach possible within the constraints imposed.

In most institutions, even where dissection is carried out, students are likely to experience more than one educational method as they pass through the anatomical components of their undergraduate careers. In Glasgow, medical students learn gross anatomy by a mixture of dissection, prosection and other methods. At Glasgow too, the dental students dissect the head and neck in their third year but have instruction in the general body systems in the earlier years that utilise prosections. This arrangement conforms very much to the ideal of William Hunter, the great pioneer of anatomical teaching, when in the late 1740s he offered students the opportunity to dissect for themselves, what he called dissection in the same manner as at Paris (Marshall, 1970). For this to be effective, he recommended a preliminary course in which the anatomy was demonstrated by the lecturer. Unless students are undertaking some sort of special study module, few courses will be able to make proper use of dissection manuals and most institutions conducting dissection courses will guide the students either by predissection talks giving instructions or by locally produced guidance notes.

INTEGRATED COURSES

Some curricula are defined as "integrated", but integration can mean different things. For example, basic medical sciences such as anatomy, physiology and biochemistry could be integrated. Maybe there could be integration of basic sciences and clinical teaching, such as by including anatomy sessions in the later parts of the course to refresh student knowledge and enhance appreciation of the anatomical basis of the diagnosis and treatment of certain conditions. Integration of subjects within a particular year of a course is sometimes called "horizontal integration" and integration of basic sciences and clinical teaching is referred to as "vertical integration", traditional courses in the past having generally placed education in the basic sciences before clinical tuition (Dahle et al., 2002; Spencer et al., 2008). The concept of revisiting topics in more depth or in more applied ways as the years of a course unfold is also a form of "vertical integration", sometimes referred to as a "spiral curriculum" (Harden and Stamper, 1999). Depending on the form of the curriculum, it may be possible to offer dissection or prosection courses as described above. Where the integration is one of dovetailing with clinical teaching, it may be that prosection-based practicals would best serve the students' needs.

COURSES IN "CLINICAL ANATOMY"

Some medical schools offer courses in "clinical anatomy" by which is meant a course that is based around the anatomy that underpins particular clinical situations either in their presentation and diagnosis or in their treatment (Boon et al., 2002). Delivery of anatomy to a class only by means of a clinical anatomy course (Rizzolo et al., 2006) would result, I suggest, in students acquiring a disjointed picture of the human form and in them struggling to integrate the information presented into a general picture of the structure and function of the human body. A clinical anatomy course, however, could be a very useful refresher for students undertaking later, more clinical, elements of the course. My opinion is that clinical anatomy courses are only appropriate for students who have already undertaken a more comprehensive anatomy course. A medical course where the only anatomical input is clinical anatomy risks being a light and inadequate affair. It all depends, of course, on what is meant by clinical anatomy. It could be argued that all anatomy is clinical anatomy, but the general meaning is the anatomy specific to particular clinical issues.

Pre-Dissection/Pre-Prosection Talks

Unless the course is a full dissection programme and can use a dissection manual or has some other workbook written specifically for it, a pre-dissection or pre-prosection talk is important to introduce the concepts, define the remit of the practical session and to deliver the anatomical knowledge at the appropriate level. There is no need for a full lecture and about 40 minutes generally suffices. Anatomy is a visual subject and students require some sort of diagram that they can take away that summarises the anatomy to be seen in the ensuing dissection or prosection exercise. This is helpful to i) illustrate their notes, and ii) to assist in the identification of structures during the practical exercise. My method is generally to construct a diagram of the region being dissected using a blackboard, a "chalk and talk" presentation. It works well for head and neck and for abdomen, my main dissection courses. There is a learning process in the exercise of drawing the diagram as its successful completion demands appraisal of relative sizes of anatomical features and their three-dimensional relationships. The trick in leading the class is to build up the diagram gradually which allows anatomical terms to be defined, anatomical structures to be described and their relationships, functions and clinical importance explained as the talk runs along. The lecturer, of course, has to have a clear idea of the final appearance of the diagram, the features to be presented and the sequence in which they should be added. The steps to produce the drawing also need to fit with a logical flow to the talk. Nowadays many university lecture theatres have done away with blackboards and it can be difficult to find a venue that has one, or one that is sufficiently maintained to accept the chalk! A pencil and crayon diagram using a visualiser is another option but, personally, I find that the end product is not quite as satisfying as with blackboard and chalk. Computerised drawing aids are becoming more sophisticated but I have no experience with them. The beauty of a chalk-based technique is that small slips can readily be covered with another layer of chalk and it is easy to apply the colour more or less heavily to give a three-dimensional appearance as required. It can be difficult to convince administrators of the need for the simple facility of a blackboard. Anatomists are likely to be left feeling that the skill and insight they possess to be able to create such pictures is not valued, even though they might be much appreciated by the students. Blackboard drawing is regarded by some as an old-fashioned and dusty method. There is little thought of the worth of these skills, even in days when there is much appreciation of art and interest in traditional creative techniques. In places frequented by tourists, for example, it is common to see street artists making magnificent chalk art works on pavements and visitors showing their appreciation. No apology is to be made for what others might perceive as old-fashioned methods. As I travel to meetings and speak to consultants who received their anatomical teaching from the generation of anatomists who also taught me, it is usually the carefully constructed chalk drawings that they recall with admiration.

Diagrams on Moodle

In response to student demand for Moodle-based material, last session and this for the third year dental class (BDS3) I have experimented with placing the blackboard diagrams on Moodle. In recent years, a number of students had asked if they could take photographs of the board illustrations on their mobile phones. Last year a decision was taken to formalise this. In this exercise I must acknowledge the help of the course co-ordinator for BDS3 who, through the auspices of the staff-student liaison committee, appointed a class photographer to take the photographs and transmit them to her. The course co-ordinator very kindly did the final preparation. She was very mindful of my ownership of the illustrations and was insistent that "Copyright S. McDonald" appeared faintly across the illustrations that were then made available to the students through their year portal to the Dental School Moodle site. Feedback on this venture indicates that the drawings have been appreciated. Following similar drawings in the recent medical class, a suggestion has been received that a DVD is produced of the drawing being built up! This is an idea with which computer technology may be able to assist.

Dissection Courses

Studying gross anatomy by dissection is probably largely confined to medical, dental and science students. Full formal dissection courses as envisaged 50 years ago by Romanes in Cunningham's Manuals of Practical Anatomy (Romanes, 1966) or by Zuckerman in his New System of Anatomy (Zuckerman, 1961) must be practised rarely for professional courses, although they may be for certain B. Sc. and Masters courses and special study modules in Anatomy. These manuals were designed for courses extending over 200 – 300 hours (Zuckerman, 1981), much more than is now available for anatomical education on professional courses. Students, however, may benefit from a shorter dissection programme, albeit less detailed than those taken by their counterparts of half a century ago.

An issue precluding dissection is often said to be one of cadaver supply, but whether this is truly so in the Western World, or an excuse to phase out dissection is more debateable. It is certainly a very real issue in Muslim countries. In the West of Scotland, Glasgow is blessed with a very supportive and public-spirited local community and in a typical year might accept around 60 bequests with half of these being allocated to the 30 dissection tables for undergraduate use and the others used for postgraduate courses. A carefully considered media appeal for donors can increase numbers, although clearly there is a delay before this can be effected.

Donor uplift, embalming, technical support and final disposal of the remains make dissection a costly process and care needs to be taken to maximise the use of dissecting room material. Judicious budgeting of undergraduate monies and careful costing of postgraduate professional courses in gross anatomy can offset dissecting room running costs and make dissecting rooms financially viable.

As to how many students should be allocated to each cadaver, my view is that the optimum number is six. Any more than that and students start to have difficulty seeing and studying features together. In practice, however, the balance of student numbers, cadaver availability, room size and possibly costs may mean that it is not possible to achieve this optimum. For the medical class in Glasgow, it is common to have 8 - 9 students per cadaver. There are 30 cadavers in the dissection room and about 250 students. In the present session the University of Glasgow is introducing more dissection and we are experimenting, with practicals with two parts, dissection and some other activity, such as histology. In the first half of the class, one group will dissect while the other group goes to the accompanying activity. In the second half of the practical session, the groups will swap over, and the group that started with the other activity will study the dissections made by their colleagues and, carry out further dissection if appropriate. At the next practical class, the group that previously was the second into the dissection room will go there first, and so on.

The above example paired dissection and microscopy but, depending on the course, it may be desirable to integrate gross anatomy with other disciplines. For example, there could be integration of anatomy and radiology. In recent years at Glasgow, successful collaborations with gross anatomy have been made with histology, radiology, physiology, echocardiography, ultrasonography, pathology, bone metabolism, and clinical skills.

In the classes with a dissection half and a histology half, we found that 75 minutes in each activity worked well. One hour for each would have been too short. Clearly the people in charge of the two parts of the class need to synchronise watches and define the change-over time. We are fortunate in having a dissecting room with two doors and we found that the smoothness of the change-over was improved if the students exiting from the dissecting room used one door and the incoming students used the other. Our experience was that the dissection by the first group was not so radical that the second group could not understand the anatomy displayed and there was always more that could usefully be achieved by the second group. It is always a good idea to have some additional exercise that could be carried out by the second group to extend the dissection should most of the initial plan have been achieved. The chart below shows the plan for running the class three times in the course of a day: for Groups 1 and 2, Groups 3 and 4, and Groups 5 and 6.

		Gp 1	Gp 2	Gp 3	Gp 4	Gp 5	Gp 6
Session 1	1st hr	Dissect Tables 1 - 10	Histology	Dissect Tables 11 – 20	Histology	Dissect Tables 21 – 30	Histology
	2nd hr	Histology	Examine Tables 1 - 10	Histology	Examine Tables 11 - 20	Histology	Examine Tables 21 - 30
Session 2	1st hr	Histology	Dissect Tables 1 - 10	Histology	Dissect Tables 11 - 20	Histology	Dissect Tables 21 - 30
	2nd hr	Examine Tables 1 - 10	Histology	Examine Tables 11 – 20	Histology	Examine Tables 21 - 30	Histology
Session 3	1st hr	Dissect Tables 1 - 10	Histology	Dissect Tables 11 – 20	Histology	Dissect Tables 21 - 30	Histology
	2nd hr	Histology	Examine Tables 1 - 10	Histology	Examine Tables 11 - 20	Histology	Examine Tables 21 - 30
Session 4	1st hr	Histology	Dissect Tables 1 - 10	Histology	Dissect Tables 11 - 20	Histology	Dissect Tables 21 - 30
	2nd hr	Examine Tables 1 - 10	Histology	Examine Tables 11 – 20	Histology	Examine Tables 21 - 30	Histology

TEACHING PATHOLOGY IN THE DISSECTING ROOM

It has been said that the dissecting room can be a place to learn pathology (Chun et al., 2007; Wood et al., 2010) but studying pathology as it is seen in the dissecting room seems a very haphazard experience. While this might make an interesting special study module for a small group of students it is not an appropriate way to deliver pathology teaching. The material is too unpredictable to make for an ordered and comprehensive pathology course.

How to Dissect

Most years, at staff-student liaison committees, a request comes up that students feel there should be more instruction on how to dissect. Their suggested solution is often that they would like me to give a lecture on this topic. I could imagine few things more tedious! Other than general tips, there is little that could usefully be said in advance and this is, in my opinion, best addressed as students work in their groups. In general, students tend to be too careful. When reflecting skin, however, they tend to go too deep. Here the trick is to keep the tip of the scalpel blade on the skin side of the hypodermis as the flap is raised and reflected. The adipose tissue should not be smeared but cut or picked off. Students also tend to use small forceps when they would be better with larger ones. Fine forceps tend to tear the tissues. A common difficulty is how to distinguish nerve, arteries and veins. The answer is that nerves often have a particular appearance in which the internal bundles of neurons can be seen. The veins are thin walled and generally the dark colour of the contained blood can be discerned. Arteries in contrast tend to be round in cross-section and tend to have pale, springy walls. Very often, of course, it is the location that indicates the identity. For example, the facial artery as it crosses the mandible lies anterior to the facial vein and no nerve accompanies them. At the femoral triangle, it is the relative positions of the femoral nerve, artery and vein that tend to be important in their recognition, although the colour and texture of each is also helpful along with their relationships to the femoral sheath.

Gloves

Students may wish to wear gloves in the dissecting room. At the University of Glasgow, because of the number of students using the dissecting room facility, we request students to provide their own. Most wear disposable gloves but, for many aspects of student dissection, washing up gloves are perfectly acceptable. Care has to be taken that students are not pinching gloves from other laboratories! One year the co-ordinator of a course that involved students coming for 30 dissection sessions over a semester said that it would be appropriate for gloves to be provided for the students. After I said I looked forward to receiving the 7,000 disposable gloves that would be required I heard no more about it!

Washing Facilities

It goes without saying that a dissection room needs adequate washing facilities. Constant monitoring and replenishment of soap and towels by dissecting room support staff is essential. If students feel they require nail brushes, they should provide their own. Communal nail brushes are a potential source of infection.

Prosection-Based Courses

Many courses involving gross anatomy at the University of Glasgow are delivered using prosections, pre-prepared dissections. At Glasgow, plastinated specimens are used, the technique by which gross anatomical specimens are impregnated with plastic (Von Hagens et al., 1987). These have advantages and disadvantages. The advantages are that the specimens do not require to be stored in tanks and kept moist and are dry and clean for handling. They can also be left out between classes. Disadvantages are that they are relatively rigid such that it is not possible to reflect more superficial structures to demonstrate deeper ones. They are also somewhat brittle and small structures, particularly blood vessels and ducts, may break if subjected to pressure. Special facilities are required to plastinate large specimens and at Glasgow we are restricted to relatively small parts. Many prosection-based courses utilise specimens labelled with flags and the use of pins, even fine ones, over a period of time tends to disrupt the tissues of the specimen. The preparation process requires time, space and costly equipment and reagents. There is also a disposal issue as the plastinated nature of the material means they are not biodegradable if buried and would thus require incineration, with specific issues over the possibility of toxic substances in the fumes. Disposal of plastinated specimens is an issue that requires to be addressed. At present it is a problem that is building up for the next generation to face!

The number of prosections to provide for a practical class is an important issue. As a rule of thumb, I always adopt the formula of no more than 4 students per prosection. Thus a group of 80 students requires a minimum of 20 prosections. That is not to say that at any given time students are confined to groups of four but it is simply a way of achieving a workable balance.

Attention also needs to be given to the size of the student group and the time allocated to the class. Let us suppose it is suggested that 90 minutes is allocated to a group of 120 students. The rule of thumb would indicate that 30 specimens will be required. It probably requires about four minutes at least for students to discuss a specimen amongst themselves and to identify the flagged features. This would mean that it would take at least 2 hours for the group of 120 students to work round all the specimens meaning that the class would be unworkable. By contrast a group of 80 would work round 20 specimens in 80 minutes, fitting well into the 90 minutes available. The group of 120 students thus needs either to come in a smaller groups or the amount of time should be increased. Consideration also needs to be given, however, to the number of specimens available and to the amount of material that it is desirable that students should see. For most practicals, it is reasonable for students to work round 20 – 25 specimens. Thirty specimens seems excessive and risks losing the students' interest. Thus a balance needs to be struck between the number of students taking a prosection-based practical, the time available for the exercise and the number of specimens it is appropriate for the students to examine.

Ideally, a prosection-based practical should follow a lecture or introductory talk outlining the scope of the practical and giving an overview of the theme of the session in appropriate detail. A summing up at the end is also good practice but it needs to be succinct as students often tend to be restive by the end of a morning or afternoon. After the introduction, the students are invited to work round the specimens. A satisfying prosection-based practical can be readily achieved by asking the students to identify two flagged structures, A and B, on each specimen. More items could be labelled but, for large groups, two items to identify ensures that the students keep moving round the specimens. It should be obvious, but worth stressing, that anatomical structures to be identified have to be clearly labelled. Any ambiguity will, if there is a be a spot-type examination, lead to questions over whether the specimens in the exam will be clear and indeed more serious concerns about the robustness of the examination. Sometimes for clarity, it is appropriate to phrase questions in some other format such as "Identify nerve A." or "Give primary function of cerebral cortex at B". A muscle like masseter, for example, consists of fleshy and tendinous fibres, so that asking students to identify the muscle makes the nature of the expected answer clear and removes ambiguity about whether a whole structure or one of its component parts is intended. It would also make clear for a muscle such as digastric that it is the whole muscle and not one of its bellies that is to be identified. Perhaps a prosection shows a dissection of the attachments of the diaphragm; "Identify fibrous band B." might make it clear that the medial arcuate ligament is intended rather than the diaphragm or the psoas fascia. Some anatomists may use a labelled photograph of the relevant specimens to convey the questions. Such a method, however, might encourage students to study the photographs rather than the actual specimens. The requirement for a photograph, of course, means that the photographs have to be taken and labelled in time for the class, which thus becomes labour intensive. Such a method also makes it necessary for the specimen photographed to be available on the day of the practical which might not be possible because of use of specimens for other educational activities. I well remember the 'unpleasantness' caused when one of my colleagues attempted to run a practical in this fashion when I had sent one of the specimens for use at a surgical examination! Important to the philosophy of a prosection-based practical are i) defined learning points to be gained from each specimen, ii) the means for students to learn by discussion of the specimens, and iii) time for students to check their answers and to revisit the specimens so as to learn from their mistakes. A practical where a range of prosections are simply laid out for the students to examine would be hopeless, and didactic teaching by a tutor loses the spirit of individual enquiry and understanding. Consider a format with, for example, "On this prosection of the posterior abdominal wall, identify i) aorta, ii) left renal vein, iii) left ureter, iv) right genitofemoral nerve, v) right suprarenal gland". The flaw here is that, other than by supplying a labelled photograph, there is no mechanism for students to be able to check their answers without staff help.

The means by which the questions are laid out is also worthy of consideration. One of my colleagues some years ago produced a number of perspex strips about 25 cm long and 3 cm broad. The questions are typed out, usually in 14 font with 4 lines of spacing between each. The page of questions is then cut with a paper guillotine into individual questions and attached to the perspex strips with Blu Tack. We have found this to be a simple yet excellent solution to the problem of how to lay out the questions. The questions can be readily taken in if required between practicals for different groups of students and readily laid back out. It saves attaching questions to the tables and the Blu Tack and the weight of the perspex stops paper blowing about.

One or two educational concepts run through the student experience. Prosection-based practicals should have a structure and an ambience which enables students to discuss the prosected material among themselves. This is crucial to the learning experience. The students should be encouraged to work in small groups of 2 – 4 individuals but should also be instructed to have paper and to write down what they think the answer is as individuals, not necessarily the group consensus. Before I appreciated the importance of individuals committing their answers to paper, I remember a class faring extremely well when they worked round a practice spot-type examination. When it came to the actual examination a week or so later, it was a decimation. The students had been able to come to the correct answers well as a group but were very poor at identifying the features as individuals. The group answers may even have given the students a false impression of their knowledge level.

Clearly, the answers need to be made available. I like a discreet system by which students can check their own answers without the embarrassment of being seen to have got questions wrong. The timing and mode of delivery of the answers to the questions to the students might vary with the difficulty of the practical exercise (see below). Usually the questions are ones for which the students have been relatively well prepared by the introductory talk or other previous material. In such instances, a list with the answers can be posted in the latter half of the class. The trick is to post them at a time such that the majority of the questions have been addressed by the class. Put the list up too early, it is likely that the students will refer to the answers without the educational element of thinking through the question to reach the answer. Too late and there is insufficient time for students to check where they went wrong.

Occasionally, a practical may be the vehicle by which new knowledge is to be imparted, i.e. the material is not being covered in detail in the introduction. Thus, the students are not expected to recognise many of the features at the start of the practical. In such instances, I place the answers to the individual questions, often with a brief explanation, on the back of the perspex strips that bear the questions. The answers are thus readily accessible for the students to view as soon as they have assimilated the question. I have used this approach particularly with the autonomic nervous system. An introductory talk describes the broad principles and features of the anatomy but the practical allows more of the local details to be considered. It is imperative in such a practical to tell the students right at the start that no prior knowledge is expected but that by means of the talk and consideration of the practical material they will experience a steep learning curve that will rapidly bring their knowledge to the required level. Failure to explain this to the students results in the students feeling that the questions are away above their heads and giving up at an early stage. This style of class, and probably all prosection-based practicals, would benefit from students being encouraged to bring appropriate textbooks to the class. Sometimes the questions asked might usefully include page references for recommended student texts to allow them to more readily use books or other notes to identify the items for themselves.

For staff, one of the disadvantages of a prosection demonstration is the time required to set it up and to write the questions. It is very helpful if support staff can look out the specimens in advance. A list of specimens with specimen registration numbers might be helpful here, although this might tend to make the specimens the same from year to year and lose the scope for improving the practical. It takes about 2 – 3 hours to find and flag the specimens and write the questions and answers. I generally take a dictaphone to record the questions and answers ready for typing.

Selection of Learning Objectives

Modern courses rightly insist on having learning objectives. These can be a bind to produce but are very helpful to both the staff and students in that they define the curriculum and the examination remit. They are probably best set by senior staff with a broad grasp of the subject areas. The learning objectives shown below come from a recent lecture and practical on the intestines for second year medical students. The objectives need to embrace all the parts of the subject that the staff deem to be important yet be fairly succinct. On this occasion seven objectives were set and I would be reluctant to have any more than this. Note that the objectives themselves do not need to name specific anatomical features but yet, in their general terms, still specify the knowledge base to be achieved.

By the end of the lecture and practical sessions, students should:

1. Be able to name and identify the gross components of the small and large intestines.
2. Appreciate the general positions and gross features of the small and large intestines in the abdomen and be able to distinguish them on gross specimens and on radiology.
3. Be able to identify the individual regions of the small and large intestines.
4. Be able to distinguish small and large intestines on histology and appreciate the microscopical differences between these two general regions of bowel and their functional implications.
5. Be able to describe ways in which the structure of the small intestine is specialized for absorption at the gross, light microscopical and electron microscopical levels.
6. Have an overview of the development of the intestines and the causation of common embryological anomalies.
7. Understand the general pattern of the blood supply and the lymphatic drainage of the bowel and its clinical importance.

Demonstrators

Demonstrators to answer the students' questions and to confirm the identity of anatomical structures in both dissections and prosections are essential to any practical. In dissection classes I would recommend one member of staff for every three tables. A system whereby any demonstrator can be called to any student group works well, especially where anatomical experience between staff varies. In prosection-based practical classes, the format I described above would require at least two staff for 80 students.

Anatomy demonstrators, it must be remembered, generally have relatively little experience of anatomical teaching and are apt to have difficulty separating the broad patterns and important principles from more specific detail. I well remember as a second year medical student wrestling with blue and red muscle attachments in Cunningham's Manual of Practical Anatomy, almost certainly driven by a mixture of student hysteria and inappropriate demonstrator guidance. In those days the demonstrators were surgical trainees who were probably themselves struggling with these details. It is thus important that the anatomist leading the class is of relatively broad experience and is the person responsible for the instruction and for defining the appropriate depth.

It is important that staff and students are all clear that the role of the demonstrator is to assist with recognition of anatomical features and to answer the students' questions but not, repeat NOT, to start doing didactic teaching. All it takes is for a demonstrator to start a tutorial and introduce a topic at an inappropriate level or some concept that goes beyond what the course leader intended and confusion will reign. I ban any attempts to draw diagrams in practical classes in which I have introduced the subject and provided a diagram. Again, any deviation from the original message is apt to precipitate student confusion and needless anxiety. A popular demonstrator who starts a group tutorial rather than demonstrating will draw large crowds and derail the intention of the practical by removing the student-driven enquiry. There will be grievances come the staff-student committee that students could not join such a group or hear what was being said and that staff are not distributing their time equally amongst the students. It is absolutely essential to the smooth running of the practicals that demonstrators are reliable and arrive on time. Demonstrators arriving late negates all attempts to insist on students arriving on time. Being not so far from the ages of the student groups, the demonstrators can be valuable in detecting issues requiring care and bringing these to the attention of more senior staff.

Some years ago, in my own university, a plan to have junior hospital staff come to demonstrate on a once-a-week basis, was a total failure. The system did not give continuity. When the hospital staff were only coming to a proportion of the class's practicals, the department had to staff the practicals on days when no hospital staff were available. Very often the hospital personnel phoned in at short notice to say that they were unable to attend because of having to provide sickness or holiday cover.

Going back 20 years in the United Kingdom, demonstrators were generally junior surgeons preparing for postgraduate surgical examinations. The lack of recognition by the surgical colleges of the value of demonstrator posts in surgical training along with the necessities of economy in universities led to the demise of this popular arrangement. Sources of demonstrating help vary across universities. At Glasgow, we generally employ recent graduates from the B.Sc. Anatomy course on an hourly basis. Other institutions retain more formal arrangements (Abdalla, 2011).

This session, perhaps with the difficulties of the job market, I found I had more applications for hourly-paid demonstrating posts than I had places. I counsel anatomists to take care who they employ, as someone poorly informed, or late, or lackadaisical, will only pull a course down. Senior staff also need to ensure that they are not being "used". I once stood firm against a recent graduate wishing to come and demonstrate as I was sure he was going to use his experience with me to bolster his application to enter a profession for which I felt him unsuited. Similarly, I also had a former graduate express a wish to come and demonstrate in a class, on the grounds of how much the person had enjoyed anatomy, etc. etc., but enquiry revealed that other colleagues were receiving a similar story and it was all a ruse to make a case for a part-time contract.

INTRODUCTION TO THE DISSECTING ROOM

Many articles such as that by Marks and Bertman (1980) have been written about how to introduce students emotionally to the dissecting room. It is my contention that negative experiences of dissection are hugely overstated. I generally tell the students about to dissect for the first time that I have found it helpful to reflect that I am not going to see or experience anything worse in my professional anatomical and forensic practice than my grandfather had to endure in the Trenches in the Great War. Students should be told about the source of cadavers for dissection. In the United Kingdom, the bodies are all generously donated by members of the public, generally from in the region local to the medical school. It is worth emphasising that the donors are well-informed public-spirited individuals who make an anatomical bequest specifically so that students and health care professionals can learn from their bodies to the public benefit. It is easy for misunderstandings to come in. Students may hear of people leaving their bodies to medical science and believe that donors have been misled into thinking that, by leaving their bodies to anatomy, something spectacular but unrealistic will be achieved such as a "cure for cancer". This is not the case. Donors receive appropriate correspondence explaining the use of cadaveric bequests, and are generally very well-informed individuals.

A safety policy is essential and should be issued to students either in hard copy or electronically, perhaps as part of a course information document. It is appropriate to highlight the main points particularly if a dissection course is to be run. This will include telling the students about the general safety of embalmed cadaveric material, endeavours to screen donors to remove risks such as hepatitis B and C and prion disease, and highlighting any microbiological hazards such as bacteria in the large bowel. This would also be an appropriate time to mention dress code in the dissecting room. White coats should be worn so students can keep themselves clean but also out of respect for the donors. Non-open shoes should also be mandatory in case a scalpel falls on someone's foot. Most importantly, students should be instructed not to attempt to fit scalpel blades until shown how to do this by staff. Hair-raising methods can occur if this step is omitted. Students should be instructed never to carry loose scalpels in pockets. I remember a lad years ago with a scalpel that had been projecting from a lower pocket and when he leant over a sink to wash his hands ... well, I was unable to say he had not opened his peritoneal cavity and had to refer him to Accident and Emergency for proper assessment. It is also worth emphasising the need for awareness of nearby colleagues when using scalpels. Another occasion I recall was a lad telling a story and, in making a point, bringing his scalpel down on to the dorsum of the hand of another student. I had not appreciated that the dorsal digital expansions in life are a spectacular silver colour rather than the cream colour seen in embalmed material but I am not in a hurry to see it again! All accidents, of course, should be reported to a designated staff member for insurance purposes.

Recently, HM Inspector of Anatomy for Scotland has made it mandatory for all students using gross anatomy facilities to sign a form agreeing to the terms of dissecting room practice. This certainly had the effect of sending a message of appreciation of the donors and about safety in the laboratory. Such agreements should be completed as soon as possible at the start of the course. After the course has started, it is amazing how difficult it can be to track students to get forms signed.

Embalming Methods

Since the last quarter of the 19th century, most dissecting rooms have embalmed using fluid based on formalin, mostly by perfusion. Occasionally small units have used immersion methods but organ preservation is inferior and may be poor. The difficulty in providing and maintaining a working environment with sufficiently low atmospheric levels of formaldehyde has caused some medical schools to move away from dissection courses. Although formalin, has served Anatomy well, the recognition of possible harmful effects has led to experimentation to develop other methods. Recently some anatomy units have been experimenting with various embalming protocols including the Thiel and Genelyn methods (Jaung et al., 2011; Eisma et al., 2013), both of which render tissues softer and more life-like than formalin. Some centres have introduced dissection of fresh or thawed frozen cadaveric material but this is for postgraduate use where closer resemblance to the natural state is required and, because of infection concerns, is not appropriate for undergraduates. I suggest that great respect must be paid to following recommended recipes for the embalming fluid for fear of reducing microbiological safety. Experience at the University of Dundee suggests that the Thiel method works well and has increased the versatility of the cadaveric material for different groups of users (Eisma et al., 2013). It also seems to require more hours of technical support than would be the case for formalin-fixed material and, like wet formalin-fixed material, is less amenable for prosection-based classes because of the need to prevent drying during and between practical classes. The technique gives a more life-like texture. Organs such as the lung and liver are soft and do not show the impressions of other organs as occurs with formalin-fixed material.

CCTV Presentation

In recent years, many dissecting rooms have installed close circuit televisions. This is a useful facility, although my own use is largely confined to demonstrating osteology and occasionally to showing and discussing a diagram of some key concept such as the fetal circulation. It may be useful for demonstrating smaller prosections but, it seems to me that larger dissections would require a portable rather than a fixed camera. It is my experience that the projection of a two dimensional image of a dissection on to a screen is hard for viewers to follow. Local landmarks may be difficult to discern in a two-dimensional image or be beyond the field making it hard to understand the location and orientation of the image. The uncertain magnification provided by the camera arrangement, and the two-dimensional nature of the television view, also means that students may struggle to grasp the depth and scale.

Computer screens in the dissecting room, however, is a direction in which we should be travelling to allow correlation between cadaveric anatomy and anatomy as seen in radiological, magnetic resonance and other modalities of clinical imaging. In hospital, radiological images tend to be viewed on computer screens. This, of course, loses the concept that actual radiological films were of similar size to the anatomical features represented. Computerised tomography moved away from this years ago. Student, however, need to have a grasp of the actual size of anatomical structures in order to be able to relate size-reduced images of body structures to patient surface and internal anatomy. With this proviso, computerised images for teaching radiological anatomy require development. In this respect, anatomy units are probably running behind hospital practice.

Class Register

Should a register have to be taken, I have found that this is best taken at the end of the practical session with the provision of a dedicated table for the signing-out sheets. For large classes the process of signing the register is accelerated if separate sheets are provided for quarters of the class, either as an alphabetical or some other appropriate group-based format.

The signing out sheet being at the end of the session ensures that students attend the whole class and it means that students can sign the sheets after they have washed up.

Osteological Hire

Oddly, it seems an old fashioned idea, but a skeleton is a most useful adjunct for a student taking an anatomy course. Up to about forty years ago it was common for medical students to possess their own half skeletons and for them to be sold on from one year of anatomy students to the next. A skeleton hire service may be invaluable to a class. It is probably now rare for such a facility to be available, not least because of student numbers. An understanding of the bones is key to appreciation of the three-dimensional layout of a region. Currently, the BDS students at Glasgow study head and neck anatomy over one semester and may borrow a skull on the payment of a returnable deposit of £60.00. Having ready access to a skull allows students to master more fully the osteology of the cranium and of the facial skeleton in the vicinity of the oral cavity and the courses of nerves that form the basis of oral local anaesthesia. Many plastic skulls do not show the bony canals and foramina clearly and we are fortunate at the University of Glasgow to have sufficient real skulls in the teaching collection to be able to offer this service. As to the value of the deposit, a balance needs to be struck between what the students can afford and a strong encouragement to return the skull. The deposit at the University of Glasgow has been £60.00 for several years but would be raised for next session should any of the present students fail to return their loaned skull.

Memorial Service and Book of Remembrance

Two years ago HM Inspector of Anatomy for Scotland insisted that all Scottish universities hold an annual memorial service for the next of kin of the donors. The administration of this in terms of organising the service and in issuing the invitations to the relatives is a large task and needs great care. I am pleased that the services that were held this session and last seemed to run successfully and were appreciated by the donors' families. Seating constraints in the University Chapel limited the number who could attend which was restricted to the relatives, some student groups and appropriate staff. At the time of the bequest, the anatomy secretary asks the next of kin if they would like to be invited to the memorial service. The service is held in late October, about 5 weeks after the start of the university session. Those invited are the next-of-kin of donors who died in the year from the start of the previous session to the start of the current session. This arrangement avoids relatives bereaved in the month prior to the service being invited. The next-of-kin is invited to bring up to two guests. Several students who had read in the media about the innovation clearly wished to attend and it was my impression that they were disappointed that on these occasions it was not possible. Increased student participation is clearly an issue that needs to be embraced as the University gains experience in holding the service. Although many universities have had successful annual memorial services for many years, staff in Glasgow were guarded about it. As well as concerns about the possibility of offence being caused by the way an individual student might have dressed or behaved, we were concerned about the much greater mix of religious faiths in the west of Scotland than in the east and staff were, and still are, rather divided on the appropriateness of a religious or a secular service. There was also a strong concern that, while a service might be appreciated by the recently bereaved, it might rekindle unhelpful grieving in those who had been bereaved many months earlier. General feedback, however, was that the interfaith service that was conducted by the University Chaplain with contributions by staff and former students had been appreciated. The previous practice at Glasgow of having a book of remembrance, a vehicle by which the University marked its appreciation of an anatomical bequest, has been continued. The book is updated at regular intervals by a professional calligrapher. It is displayed on a lectern at the memorial service for viewing by relatives. At other times it is retained in the Anatomy Museum and may be viewed by arrangement.

Conclusion

I hope this chapter gives a frank and honest account of my ideas on how to run a course in gross anatomy and that they are helpful to others. Anatomical courses need to be tailored to local needs and resources. Each could be regarded as an on-going experiment to enhance understanding and enjoyment of the subject. I hope I have provided practical suggestions, and some warnings, that colleagues can use to develop courses in their own institutions.

References

Abdalla, S. (2011). The role of anatomy demonstration for surgical trainees. *Ann Roy Coll Surg Engl* (Suppl) *93*, 60 – 62.

Boon, H., Meiring, J. H. & Richards, P. A. (2002). Clinical anatomy as the basis for clinical examination: development and evaluation of an introduction to clinical examination in a problem-oriented medical curriculum. *Clin Anat*, *15*, 45 – 50.

Chun, J, Theologou, T. & Ellis, H. (2007). Incidence of cardiovascular disease in the dissecting room: a valuable teaching asset. *Clin Anat*, *20*, 89 – 92.

Cottam, W. W. (1999). Adequacy of medical school gross anatomy education as perceived by certain postgraduate residency programs and anatomy course directors. *Clin Anat*, *12*, 55 – 65.

Craig, S., Tait, N., Boers, D. & McAndrew, D. (2010). Review of anatomy education in Australian and New Zealand medical schools. *ANZ J Surg*, *80*, 212 – 216.

Dahle, L. O., Brynhildsen, J., Behrbohm Fallsberg, M., Rundquist, I. & Hammar, M. (2002). Pros and cons of vertical integration between clinical medicine and basic science within a problem-based undergraduate medical curriculum: examples and experiences from Linkoping, Sweden. *Med Teach*, *24*, 280 – 285.

Eisma, R., Lamb, C. & Soames, R. W. (2013). From formalin to Thiel embalming: what changes? One anatomy department's experiences. *Clin Anat in press*.

Grodinsky, M. & Holyoke, E. A. (1938). The fascia and fascial spaces of the head, neck and adjacent regions. *Am J Anat*, *63*, 367 – 407.

Gupta, Y., Morgan, M., Singh, A. & Ellis, H. (2008). Junior doctors' knowledge of applied clinical anatomy. *Clin Anat*, 21, 334 – 338.

Harden, R. M. & Stamper, M. (1999). What is a curriculum spiral? *Med Teach*, *21*, 141 – 143.

Jaung, R., Cook, P. & Blyth, P. (2011). A comparison of embalming fluid for use in surgical workshops. *Clin Anat*, *24*, 155 – 161.

Marks, S. C. & Bertman, S. L. (1980). Experiences with learning about death and dying in the undergraduate anatomy curriculum. *J Med Educ*, *55*, 48 – 52.

Marshall, A. J. (1970). Catalogue of the Anatomical Preparations of Dr William Hunter. Glasgow: University of Glasgow.

Rizzolo, L. J., Stewart, W. B., O'Brien, M., Haims, A., Rando, W., Abrahams, J., Dunne, S., Wang, S. & Aden, M. (2006). Design principles for developing an efficient clinical anatomy course. *Med Teach*, *28*, 142 – 151.

Romanes, G. J. (1966). Cunningham's Manual of Practical Anatomy. 13th Ed. London: Oxford University Press.

Sinclair, D. C. (1965). An experiment in the teaching of anatomy. *J Med Educ*, *40*, 401 – 413.

Spencer, A. L., Brosenitsch, T., Levine, A. S. & Kanter, S. L. (2008). Back to the basic sciences: an innovative approach to teaching senior medical students how best to integrate basic science and clinical medicine. *Acad Med*, *83*, 662 – 669.

Turney, B. W. (2007). Anatomy in a modern medical curriculum. *Ann Roy Coll Surg Engl*, *89*, 104 – 107.

Von Hagens, G., Tiedmann, K. & Kriz, W. (1987). The current potential of plastination. *Anat Embryol*, *175*, 411 – 421.

Wakeley, C. (1953). The teaching of anatomy for the doctor. In Proceedings of the First World Conference on Medical Education. London: Oxford University Press. 225 - 231.

Wood, A., Struthers, K., Whiten, S., Jackson, D. & Herrington, S. (2010). Introducing gross pathology to undergraduate medical students in the dissecting room. *Anat Sci Educ*, *3*, 97 – 100.

Zuckerman, S. (1961). A New System of Anatomy. Oxford: Oxford University Press.

Zuckerman, S. (1981). A New System of Anatomy. 2nd Ed. Oxford: Oxford University Press.

In: Education in Anatomical Sciences
Editor: Paul Ganguly
ISBN: 978-1-62618-488-6

Chapter 7

ANATOMICAL EDUCATION: GROSS ANATOMY (PART 2)

K. Rajendran[*]
Department of Anatomy, Yong Loo Lin School of Medicine
National University Health System, Singapore

DEDICATION

This chapter is especially dedicated to the generations of students I have been exposed to in more than 30 years of teaching gross anatomy using approaches ranging from the early, traditional (semi-independent) dissection to supervised dissection and finally moving to prosection and multimedia. Close interaction with these students as teacher, friend, and mentor has not only enriched my life as a person but also provided me with unique opportunities to continually improve teaching based on an understanding of how they learn and the difficulties they encountered in trying to cope with gross anatomy. The emphasis on education, as well as camaraderie and cooperation among the staff members of the department, have also been of tremendous support towards continual improvement in the teaching of gross anatomy. While there is abundant literature on the teaching of gross anatomy, I can only justify this chapter by coloring it with liberal amounts of my personal experience as I keep growing as an educator.

BACKGROUND

Gross anatomy is all about what we are made of, and how we are designed for function and it has had a natural fascination for man throughout history. As depicted since early days, the body has obviously appealed to people both from the scientific as well as the artistic perspective. When we refer to Gross Anatomy in education however, it becomes more of a

[*] e-mail: k_rajendran@nuhs.edu.sg.

science, and a subject that describes and explains the structural organization of the body that forms the basis for normal function.

Arising from natural curiosity about the normal functions and dysfunctions of the body, as well as appreciation of its beauty, knowledge of gross anatomy has evolved over the millennia through opportunities offered by accidents, conflicts and perhaps cultural practices, eventually leading to proper scientific study largely through dissection.

In the medical and related professions gross anatomy lays the foundation that naturally lends the human body to be treated by complementary fields such as physiology, pathology and in particular, surgical disciplines. It is therefore easy to recognize the pivotal role of Gross Anatomy in the training of medically related professions that either need to deal with or make reference to the human body, and as one would expect, the respective curricula often commences with some form of introduction to the subject.

As the first, and critical, major subject to impact students in the relevant courses, it is therefore imperative to ensure that the subject of Gross Anatomy sparks the right kind of interest, passion and motivation in the student to continue the search for more, relevant and related knowledge and not to have a dampening effect.

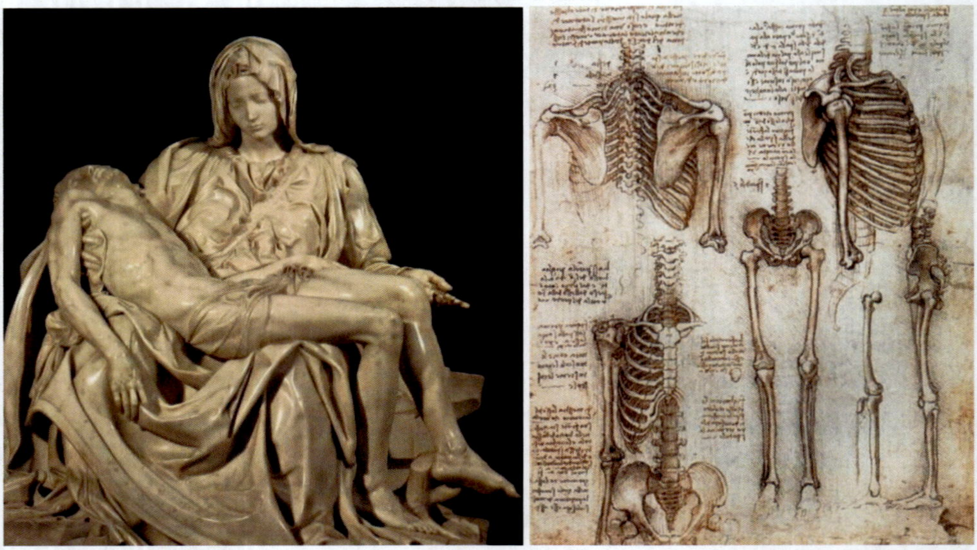

Anatomy in ART and in SCIENCE

Teaching of Anatomy

It should be emphasized that whatever the nature of a subject, the desired outcome of teaching is learning on the part of the student and not simply a grade that suggests competence. Learning a scientific subject implies understanding, the ability to explore the subject further, and to know how knowledge can be applied in practice and to raise new questions for which answers need to be found.

As Gross anatomy **'evolved'** over the years, details often overwhelmed the understanding of functional and clinical relevance. This led to the subject being very factual and less functional allowing little connection with other relevant disciplines. The result was the teaching of Anatomy in isolation and for want of opportunities for the student to learn appropriately in later years there was the inevitable over teaching of facts "just in case it was relevant at a later date". This "just in case" approach, with the attendant information overload, in turn resulted in negative perceptions about gross anatomy such as being dry, impossibly excessive in factual content and seemingly leading to little relevant application.

Fortunately, today the correlation between structure, function and clinical relevance, along with supporting technology for education can be considered to have 'matured' to a level where poor teaching becomes less than acceptable. One must however be mindful of reality where teachers, being who they are, have their own learning curve (especially as novices), personal preferences in style and approach, as well as shortcomings, elements to which the student may be subjected to in a favorable or unfavorable manner.

IMPORTANT CONSIDERATIONS FOR TEACHING

The Nature of the Subject

While we have many sensory modalities, vision is the sense that is most used to take in information about the world around us be it of a pictorial or textual nature. Gross anatomy being primarily about structure and how it relates to function therefore must qualify for a visual approach to understanding, and applying the subject. Furthermore, in the appreciation of physical structure, the tactile experience completes learning by providing information that handling of an object generates [1] such as consistency, shape, movements, etc.

Who Are the Students?

It is easy to overlook or ignore this question, but doing so could seriously hamper the approach to teaching students. Of important consideration are:

1) Are they first timers to human or any biology? This situation is becoming more prevalent locally for e.g., where the mandatory requirement for entry into medical school is Chemistry, or Physics over Biology. In situations like this, biological concepts and terminology may become daunting, and some form of basic introduction to the subject would be appropriate.
2) How mature are the students? Some students come fresh from school and often tend not to be independent learners. The approach to teaching these students could be phased in from 'spoon feeding' to one that encourages more independent learning.

Organizing the Program

Put briefly, the program should generate interest, understanding, appreciation of relevance, and as much as possible, a fairly complete picture (of human anatomy) that suits the level of learning required.

Introduction

This is easily overlooked in the midst of an extensive, detailed study of the human body but a good introduction should not be taken lightly. It should at least cover the following elements bearing in mind that many of the students could be total novices to the subject:

a) A clear explanation as to what the nature of gross anatomy is.
b) A reassurance that the language of anatomy will become second nature with usage and can be learnt 'meaningfully'.
c) Why it (gross anatomy) is being learned and how it could be used for the intended course of study i.e., medicine, dentistry, pharmacy, etc.
d) What approaches are being used for learning the subject and what is the value of each method?
e) The importance of the basic principles and concepts in making the learning meaningful.
f) The need for constant reinforcement in such a vast subject.
g) How not to run into potential pitfalls, such as relying on memory, overlooking basic concepts and not keeping up with the program.

This will go some way towards reassuring the new student that the subject is not as intimidating as they may have anticipated.

Approach to the Dispensation of Knowledge

Having figured out what is to be done, it is appropriate to examine how best it should be done, i.e., the actual education process or what the student is going to experience in the journey through gross anatomy.

1) The Big Picture

If there is something students really want from teachers of Gross Anatomy, it is evident most often in their feedback comments that say "Thank you for the Big Picture". This is perhaps what many textbooks often overlook in their preoccupation with the factual details. This is a reminder that it is easy to get lost in the 'Jungle' that gross anatomy can be.

It is prudent that for any topic, the organization should be first presented in its simplest form. For example, the entire body could be seen as comprising Somatic and Visceral components, or, areas of the body could have major components organized into layers, front to back, or superficial to deep.

To elaborate a little, in the abdomen for example, the wall, and organs could be presented as the most basic components in the form of external, somatic, and internal visceral components. The student should have little problem conceptualizing this. The viscera could then be shown as being arranged in layers much as in an open book where a single center page represents the gastrointestinal system in its simplest form while pairs of pages alongside the center page represent, from anterior to posterior, other systemic components such as gonadal (organ/vessel), followed by urinary (including kidneys and ureters) and finally the posterior abdominal wall which is fairly symmetrical in design on either side of the midline. This goes a long way toward helping the student to mentally prepare for the details that follow under each category/system.

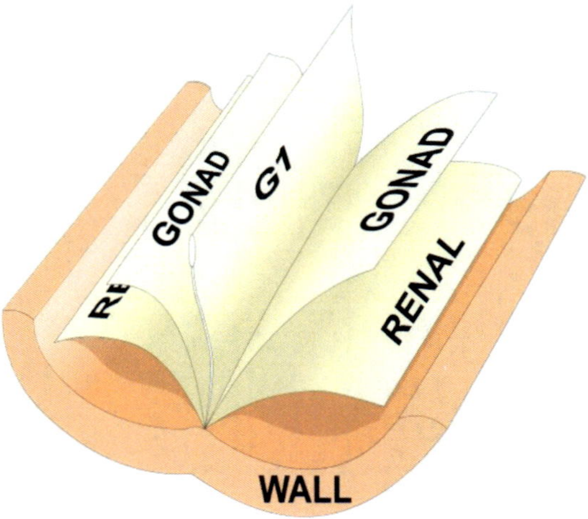

Familiar concept of "open book" may help visualize the organization of abdominal anatomy.

2) Important Concepts

Concepts often repeat throughout the body and should be given early to spare the student the frustration of repeated learning. For example with reference to body cavities, there is the concept of parietal and visceral which is applicable to the pericardial, peritoneal and pleural cavities.

Other items which benefit from a good conceptual understanding include: any skeletal muscle, its organization that allows proper function i.e., the muscle belly, and tendinous attachments to different bones that allow movements, design of bones for muscle attachments, i.e., tubercles, trochanters, tuberosities, and the basic structure of synovial joints and how articular surface design could influence movements. These prepare students to enter 'new' territory with confidence and interest and little or no apprehension.

Other often confused topics, where clear concepts are critical, include the meningeal spaces where proper understanding is critical to knowing the sources of bleeding, the possible complications arising from hemorrhage, and the impact on proper diagnosis and treatment.

3) Principles

In general, principles underlie the 'operation' of various body components and again form an invaluable foundation for the understanding of and description of anatomical entities/components. Joints, for instance, behave on the basis of articular contours, which in turn define axes for movements and planes along which the movements take place. It is then possible to understand the location and orientation of muscles that produce the various movements at any synovial joint. In taking this approach a student who understands the shoulder joint will be able to figure out how a fan shaped muscle, like the pectoralis major, is able to help produce adduction, flexion and medial rotation.

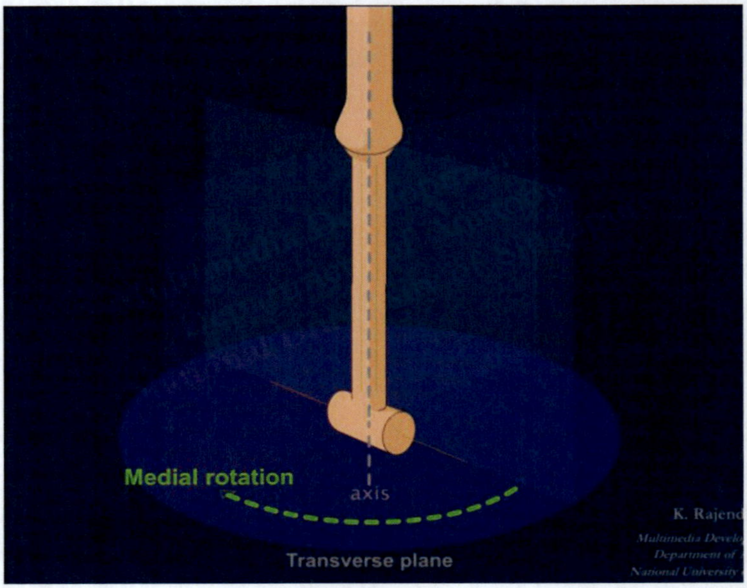

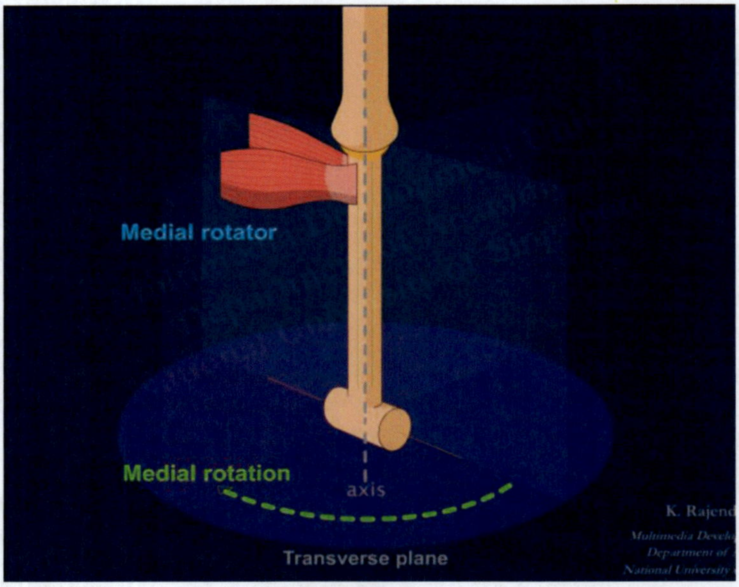

Animated concept on "ball and socket" joint illustrates the basis for movements and muscle action.

Principles could also underlie the approach to learning structures that practically serve the entire body like nerves and vessels. Looking at any nerve or vessel, a student should immediately be able to ask basic questions like origin, termination, distribution, composition (nerves), collateral circulation (vessels) and effects of compromise or injury.

4) Ensuring a Basic Knowledge

What exactly is basic knowledge? This should be the kind of knowledge that provides the foundations that allow for further advancement of the student in learning the science in a systematic and progressive manner. Perhaps as an example, even at the simplest level of learning the subject a student should understand the significance of categorizing the body into simple but functional major blocks such as somatic and visceral components. For the (lay) beginner the somatic part, which is external, is essentially 'the body'. This external component is indeed an interesting, and functional introduction to the musculoskeletal system that we use to interact with the world around us. The visceral component, made up of various other systems, can then be featured as the internal parts that are functionally organized to help sustain the somatic, external body. Progression in learning then leads meaningfully to the different systems and how they work and interact towards maintaining a healthy homeostasis.

5) Visualising Gross Anatomy

There are few subjects that lend themselves as well to a visual approach [2] to learning, as indeed does, gross anatomy. And there are few better ways to impart visual understanding than to handle a three-dimensional structure, especially with opportunities to see how complex structures come together into a functional whole.

a) Using Cadaveric Specimens to Teach

One may wonder why in today's world of virtual reality [3] with a huge resource of IT material, one should resort to the inconvenience of using cadavers to teach gross anatomy. There are potent reasons why cadavers are a highly valuable educational resource if available:

1) Human anatomy is about the real human body [4] and not virtual. The real body brings home the reality regarding variability among people e.g., in terms of age, sex, size, ethnicity, disease etc., which is what the student is going to encounter in the real world of patients. There is also a better sense of size and proportion when handling the real body as opposed to viewing IT material. Handling during study, either through dissection or prosection, gives a realistic experience regarding the composition, consistency, organization, and exposure to a multisensory input that is closer to a more complete experience.
2) One cannot overstate the intangible values imparted by having the opportunity to experience the selfless generosity of others that has made cadaveric study possible. Recognizing the cadaver as the body of a person and the necessary respect this engenders helps influence the right attitudes, in due course, towards patients. Instilling and reinforcing a sense of respect for the human being and body is an opportunity not to be missed or delayed. A ceremony accompanying the first exposure to the body and a closing ceremony are becoming accepted as a desirable practice that is quickly gaining popularity.

So, Is Dissection [5] Better Than Prosection?

Dissection - Where the human cadaver is a limited resource, its use must be optimized and dissection (as opposed to prosection) may not be the preferred mode of study since it requires familiarity (of anatomy) to be done properly. Proper, step-by-step video guidance and prior familiarization with anatomical structures, at least conceptually, should minimize 'wasting' bodies. Dissection then is perhaps best reserved for senior students who, having already studied anatomy through prosection studies and seen patients return out of special interest either in anatomy itself or in surgical disciplines. A proper dissection study takes time and should not be rushed within a compressed course.

Prosection study - This may be seen as robbing the student of a hands-on experience, but it does have advantages. Done by experienced dissectors, therefore giving a better demonstration of structural details, prosections need to be planned well such that every student has the opportunity to view anatomical structures in different states of dissection, i.e., and to appreciate superficial and intact structures before visualizing the effect of further dissection that reveals other details such as blood vessels and nerves. This has been possible by providing each group of students with at least two prosected cadavers where one shows external and internal structures relatively intact i.e., unseparated muscles, or undissected but accessible visceral structures. The second cadaver shows a finer dissection to reveal hidden details. Supplementation with further specimens with special exposures completes the need. It must however, be accepted that some essential details such as subcutaneous tissue and fasciae may be less accessible for study. The notable advantage of prosections is the possibility of customized exposures to facilitate the understanding of organization and relations, and the chance to revisit the body any number of times, unlike the dissected body that often dwindles away over the weeks or months as structures are systematically removed in the process of complete dissection.

Regional and Systemic Approach

Is one better than the other? The fact that they have both been around for a long time suggests otherwise. The health sciences tend to present normal functioning of the body often (along with pathology) leading to abnormal function. Systems usually correlate better with function/Physiology and function is something every eager student can't wait to be introduced to. In general therefore an overview of systemic anatomy could precede regional study without having adverse effects (of being 'fragmented') on the learning process.

Regional study however has been the popular method of study, perhaps dictated by the needs of the dissection process, which has been the traditional approach to learning for centuries. In many ways regional study also serves the needs of clinical practice most of the time, where a clinician examines or carries out procedures on defined areas (regions) and not systems of the body. The major setback in this approach is the limited scope of learning within defined boundaries and uncertainty beyond the boundaries of the region [6] much like selected pieces of a jig saw puzzle without the 'big picture'.

Systems therefore provide the continuity between regions and this, together with a good overview, should allow the student to navigate the body between areas (regions), a practice that is becoming all so prevalent with the popularity of endoscopic procedures that take a surgeon beyond familiar boundaries and requires the radiologist to scroll through sectional anatomy up and down body regions.

For basic courses the systemic approach can be the primary approach, combined with regions and examples of applications in order to elicit a lot more interest.

More Ways to Help Visualize

Using *chalk on a board* can produce excellent presentations where simple line drawings create a concept and progressively develop it to relate the full picture of the topic. This approach also allows the audience to follow the development of a theme in real time, and, if the situation permits, interrupt exactly where difficulties are encountered. So, as simple as this may be (to experienced teachers) it achieves the goal of educating the student on topics that lend themselves to be treated this way. This is a case of a presentation that follows the intent of the teacher but also leaves no room for major errors, as drawing is the surest way to demonstrate gaps in knowledge.

At the other extreme, there are the overly complex images (from standard resources) with distracting details that are not always helpful in getting the attention of the audience on the structure(s) being referred to.

b) Multimedia

We are at an age when multimedia has become a pervasive presence and is here to stay. Fortunately it is potentially an ally that will enhance the teaching of gross anatomy [7] by playing a complementary role in any environment where gross anatomy is being taught. Just as prosection studies may need to be customized to serve the intentions of the teaching program, experience seems to suggest that in spite of the huge wealth of multimedia material available, perhaps the best multimedia should support the individual preference of the teacher. Therefore short of choosing carefully, or customizing existing multimedia, it's perhaps best to generate one's own, in line with one's thinking on how teaching/learning should take place. They lend themselves well to original thinking and encourage the creative element in the teacher, thus maintaining enthusiasm for continual improvement in the way gross anatomy is taught. There are several possibilities for use of multimedia.

1) Graphics - To provide the necessary visual accompaniment for any of the simpler topics being presented. Labeling preferably should not be extensive (to avoid distraction) and, ideally should appear when the specific component is being made reference to.

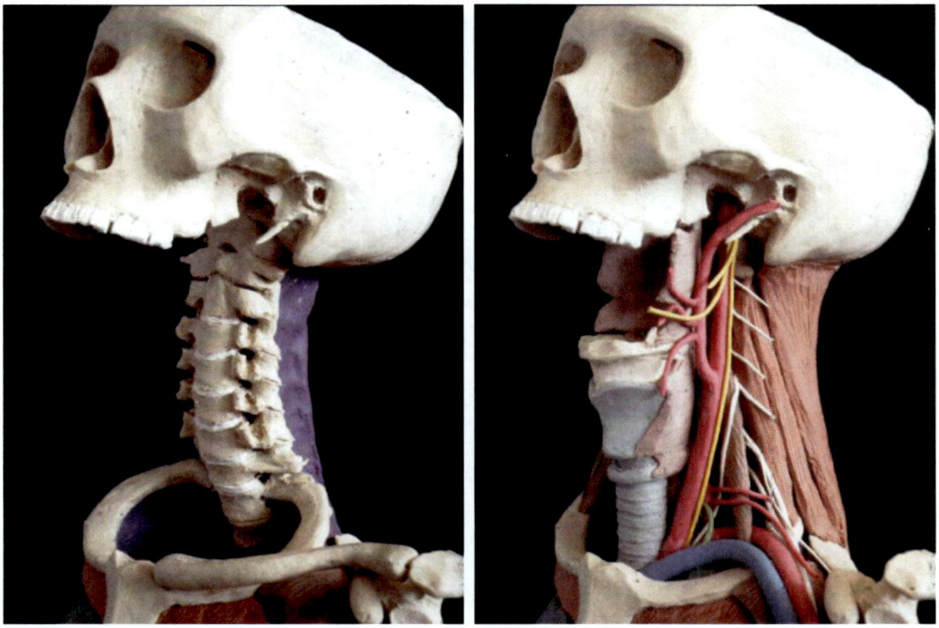

Uncluttered presentation, in selected steps, helps focus and visualisation.

2) *Simple interactivity* - This could be in the form of 'adding' or 'removing' structural components [8]. This is highly desirable in dealing with structural complexity enabling gradual 'build up' of details from something simpler.
3) Free assembling/disassembling - Ideal for complex areas of the body that offers several ways of completing the presentation of all the anatomical components. This program allows for 'assembling' the regions [9] starting from any of the components, thus giving flexibility in approach and learning about organization from different perspectives. This is not dissimilar to what a child does when he/she dismantles a toy thereby intentionally reducing complexity and then attempting to put them back together, thereby, understanding how they come together. The next natural step is the attempt to understand function.

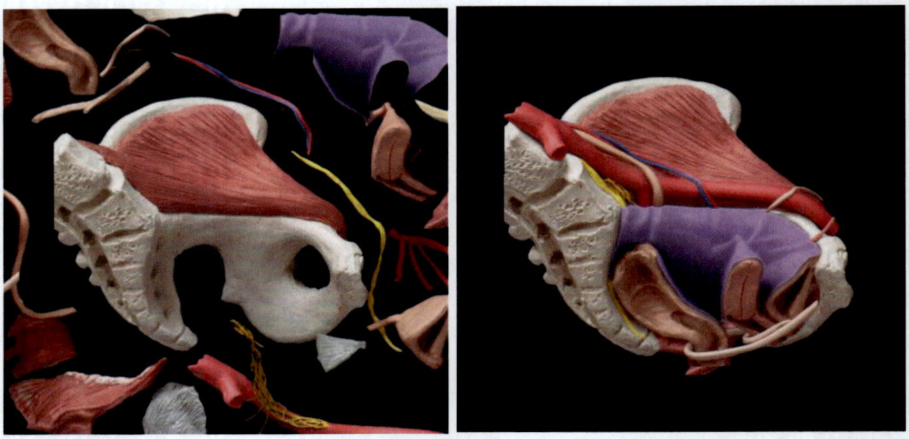

Interactive "assembling" program for better understanding of structural relations.

4) Animations - Achievable through the use of flash or dissolve programs [10] this facility truly adds a new dimension to presentations, firstly allowing changes to be shown during the developmental process, demonstrating movements, and not least, generating necessary attention on the part of students.

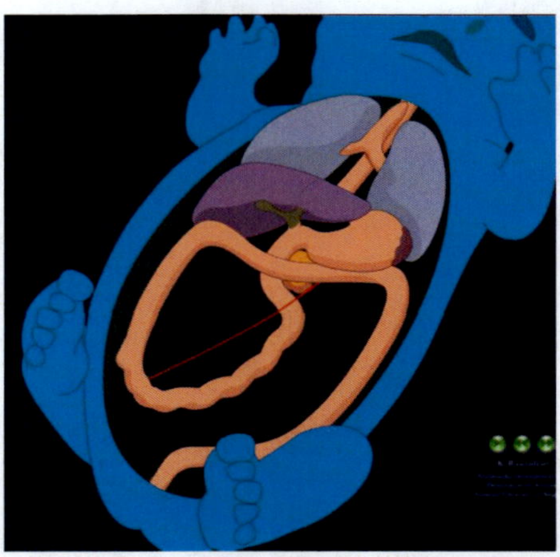

Potentially complex normal anatomy of the gastrointestinal system simplified using animated embryology.

5) Videos - Extremely useful for showing procedural examples of dissection, clinical conditions and procedures, which help add a dimension of relevance as well as interest on the part of the student.

6) Making Reference to Subdisciplines

Embryology

As a subdiscipline of Anatomy, embryology is daunting enough for the unique challenges it offers to the novice learner. However a proper understanding of some aspects of gross anatomy, such as seemingly 'strange' morphological features, requires appropriate reference to the developmental process. This need not be detailed but sufficient to make selected gross anatomical 'mysteries' meaningful and learnable. Done in proper measure, this exercise again adds clarification to otherwise confusing gross anatomical descriptions.

Examples include the internal feature of the heart, in particular, the inter-atrial septum, the mesenteries, particularly those of the stomach, e.g., greater omentum. Retroperitoneal and intraperitoneal structures add to the list.

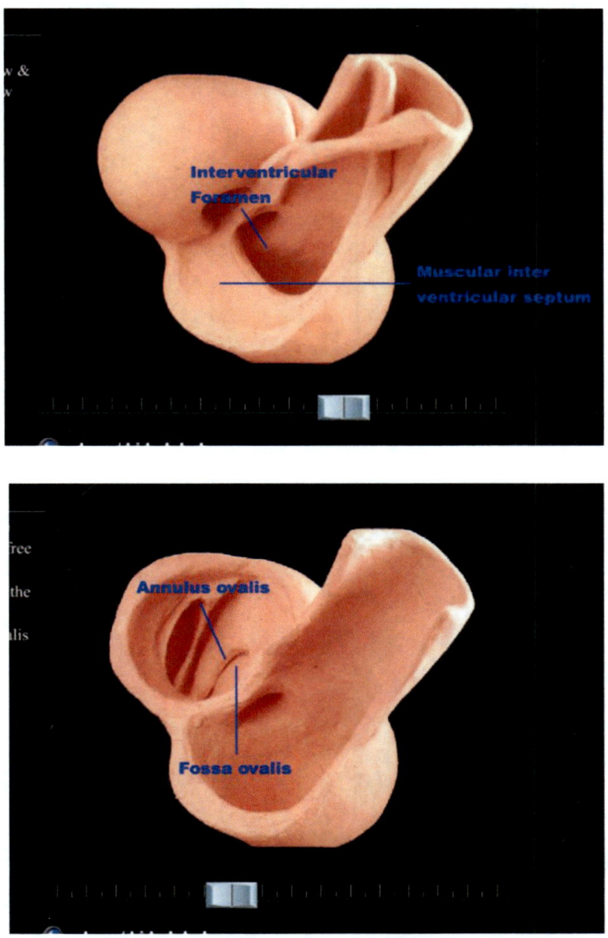

"Mystery" of normal anatomical features (like fossa ovalis) of the heart rationalized by watching developmental process.

Variations to the normal gross anatomical features, even minor ones, are often explained with a simple reference to embryology, e.g., aberrant renal vessels, maldescended testis, contents of the inguinal canal (processus vaginalis in particular).

Conversely, functional anatomy e.g., of mastication and swallowing is helpful in understanding the otherwise complex embryological topics such as the pharyngeal or branchial apparatus. Biting, chewing and swallowing, when understood functionally through study of gross anatomy point the way to remembering the significance of the nerves of the various branchial/pharyngeal arches which in fact form the surrounds of the future upper respiratory and alimentary systems.

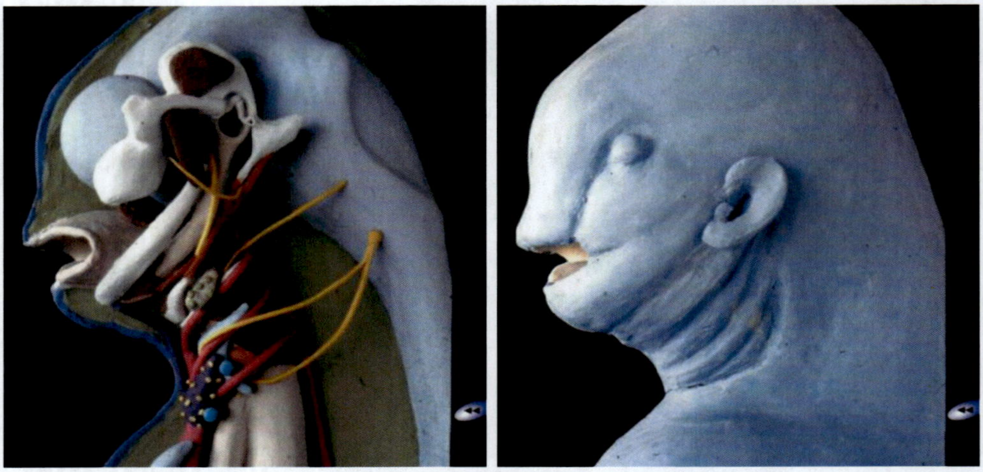

Appreciation of functional anatomy may aid in the learning of embryology.

Similarly, reference may be made to histology and physiology in a functional manner.

7) Using Approximations

Using precise details cannot only be exasperating to students, but are unnecessary as 'precise' numbers, e.g., in relation to measurements averages are often used.

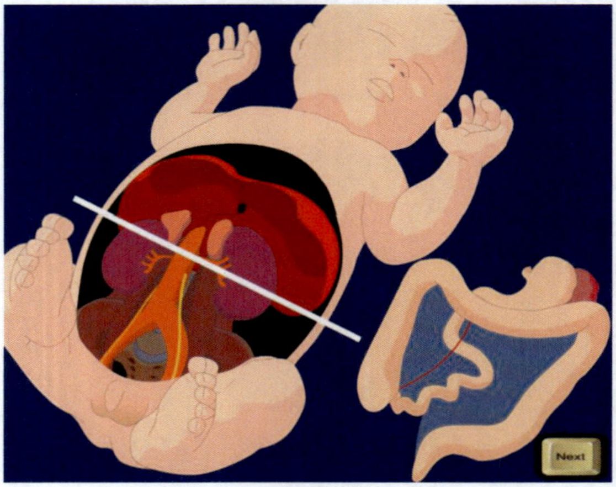

Transpyloric plane may be used as an approximation to facilitate better visualization of organization and relations of intra abdominal structures.

Proportions based on obvious (bony and other) landmarks are often more helpful and less intimidating. The mid-clavicular line is one example when used to locate the apex of the heart (instead of remembering how many centimeters/inches it is from the midline). Similarly, the transpyloric plane can be used to provide approximate but useful information regarding the organization of many of the components in the abdomen. Details can follow appropriately once structures have been approximately put together in relation to the plane i.e. those above, and those below, and in the correct order, front to back (in systems).

8) The Element of Interest

It is the hope of anyone taking up education in gross anatomy that the student wishes to learn the subject out of interest in the structure and workings of the human body. This may not be the case when gross anatomy needs to be offered as a module intended to meet the requirements of a specific course.

Paradoxically, while interest in knowing about the human body may be seen to be almost instinctive, the challenge often seems to be to sustain this interest or to further encourage it to stimulate higher levels of learning and pursuing it as a science.

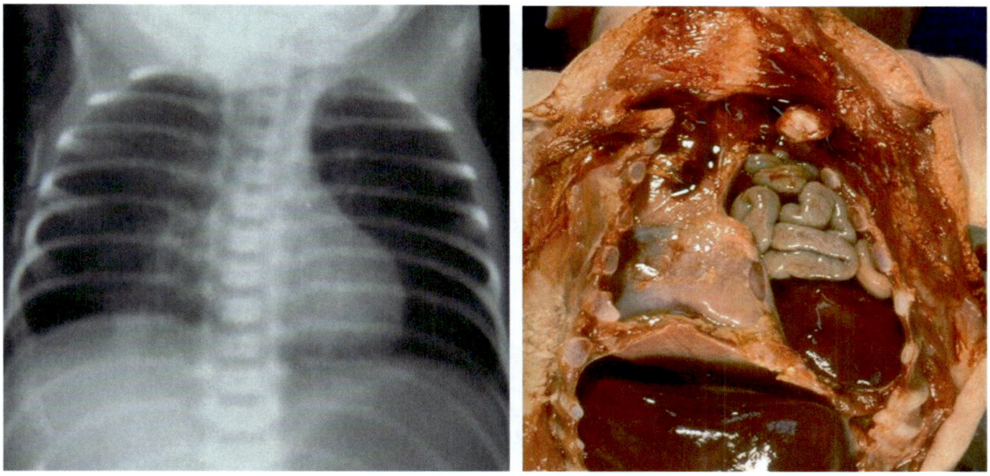

Clinical examples heighten impact and relevance.

Where does the fault lie? One cannot blame the student, the novice, when in fact it is the educators who should have helped generate the right impact. Most of us look at ourselves as anatomical beings and naturally wish to know what we are made of, what makes us tick and what might go wrong. Factors that conspire to dampen the student's interest include:

1) Failure to introduce the human body as a functionally integrated system is a formula for improper orientation right from the beginning. It robs the student of wanting to go on to other areas to explore his interest further.
2) Lack of functional connection to structure, especially in the way of everyday examples and clinical applications. This is the fuel that sustains the fire of anatomical knowledge.
3) Excess of detailed information, even if relevant, can be intimidating as this can be self-learned and may be misinterpreted as relevant to assessments.

4) Not making reference to relevant, and common, problems and how to solve them using the science of anatomy.

Use of Mnemonics

There's nothing worse than the horror of forgetting a mnemonic. It is difficult to discourage this common approach to remembering difficult subject matter in anatomy, whether it has to do with branches, relations, etc. Indeed mnemonics could range from the ridiculous to the obscene but seem to have served their purpose throughout the history of 'learning' anatomy. As a science however, it's perhaps best, whenever possible, to approach the subject through the fundamental principles, appropriate visual input and reinforcement from time to time. One is perhaps less likely to make the kind of serious errors that could result if a mnemonic is mixed up or forgotten. The Zebra Burped Munching Cornflakes can help remember the major branches of the facial nerve as Temporal, Zygomatic, Buccal, Mandibular and Cervical provided one remembered that Zebras ate Cornflakes or actually burped! Scientific subjects should have a sensible scientific approach, with or without the added 'fun' element of mnemonics, which perhaps best suit the 'crash courses' in anatomy. Whatever initiated resorting mnemonics should ultimately result in proper learning.

Logical Anatomy

The human body is undoubtedly put together in a very sensible way to allow optimal functioning. Where possible therefore, the logic of a situation should be made clear. Why, for instance, are somatic senses so precise compared to visceral sensations?

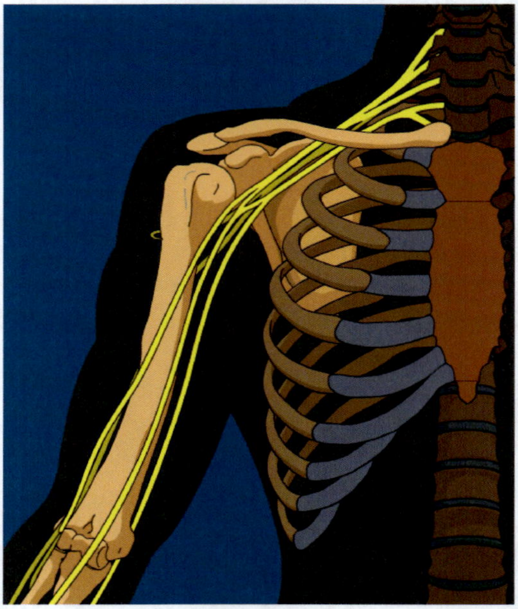

Use of the patterns (M) in the brachial plexus to simplify orientation of nerves in the arm.

Largely because the precision is helpful in dealing with whatever is causing it, unlike in a visceral situation. If the M formation is used in reference to the brachial plexus to identify nerves supplying the front of the arm, the same M formation could be used to explain the distribution of the main nerves just by projecting them downwards and laterally from the M, which results in the musculocutaneous occupying the highest position (arm) and the remaining two retaining the original relationship in the M with the median nerve lateral to the ulnar. Or, reasoning for the naming of the radial ulnar and median nerves. The possibility is the terms originated in relation to the forearm where the radial nerve is on the radial side (radius being the bone that behaves like the radius of a circle when pronating/supinating) with the median (middle) nerve in the mid-position and the ulnar nerve is closest to the ulna.

Is There a Correct Flow of Topics in Gross Anatomy?

The choice and progression of topics can have a significant bearing on the students' experience. Students in general should get on to an optimistic start, with interest, enthusiasm, and an opportunity to pick up principles and concepts in gross anatomy so that later topics, though relatively more difficult, need not be so challenging.

Musculo Skeletal System (Traditionally Covering Limbs and Spine)

Interestingly the external, somatic body is the component that attracts not only most beginners to anatomy, but also those who wish to understand the subject to better render it in artistic works.

This component is of course the musculoskeletal system; the human component that permits us to do things in the world. It is also the system first exposed in a dissection or prosection study of the human body.

Because this system has an organization that is relatively simple, and easily related to function, with fewer component parts, and introduces to (nervous) control mechanisms, it is good to start with the musculoskeletal system. Students should also feel less intimidated. In fact it is a system that in our experience consistently receives excellent student feedback regarding their learning experience.

Trunk

Moving to the part of the body containing most of the internal organs, again we could identify progression in relative comfort from simpler to more complex challenges.

The first obvious additional feature in the trunk is that the outer part, the wall, is somatic in nature. The student therefore has little need for anxiety as the general concepts and principles regarding the musculoskeletal system have already been understood while studying the limbs. The new structures are the viscera with associated vessels and the visceral or autonomic nervous system.

Thorax

Doing this as the first part of the trunk has the benefit of dealing with critically important systems, the cardiovascular and respiratory, in preparation for the coordinated study of physiology where these systems are usually covered early in the program. The arrangement of viscera is also relatively simpler and there is an early opportunity for introduction to life saving measures such as cardio pulmonary resuscitation.

Abdomen

Here, the organization of the viscera gets more complicated and relations can be a challenge. However the concept of the midline origin of the gastrointestinal system (with subsequent alteration in disposition) is useful in appreciating the otherwise seeming complexity in the abdomen.

Pelvis and Perineum

In some ways structural visualization and functional correlations are more challenging. Coincidentally this is also where most abdominal viscera terminate at their outlets, thereby requiring the student to understand the structural mechanisms behind the control of evacuation including parturition.

Head and Neck/Neuroscience

Most teachers of anatomy generally accept the increased complexity here, compounded by the cranial cavity, brain and cranial nerves. There are also notable advantages in learning this component last. The basics acquired from learning the trunk can be applied here. The neck for instance has a muscular 'wall' (used for movements, like the abdomen), a posterior wall very much like that of the abdomen where the psoas with nerves emerging adjacent to it are not much unlike the scalenus anterior and the spinal nerves forming the cervical and brachial plexuses. Within the 'cavity' of the neck are the viscera in a simple anterior to posterior arrangement.

The head may be best seen as a cranium containing brain cranio dorsally with respiratory, and gastrointestinal components applied close to it ventro caudally. Other structures seemingly adding to complexity are the accessory structures that essentially support visceral functions such as mastication.

Having understood the somatic and visceral nervous systems at a simpler level, the study of the brain allows for the understanding of higher levels of integration and control.

This sequencing therefore allows a relatively effortless and progressive encounter with otherwise potentially confounding gross anatomy.

Answering the Call to Full Integration

There have been recent trends toward an integrated approach towards learning all disciplines as an integral part of what one might refer to as human biology, thus mandating a necessary connection across relevant disciplines. This approach is laudable but has its challenges and setbacks. Disciplines are indeed connected, so the intent should be to allow the student to see where it may be connectable, and not to intentionally connect to everything it may have some relevance to. It's always useful to establish where the links are e.g., between gross anatomy and histology/pathology, or gross anatomy and surgical procedures. These links should also be matched to the level of study of the student so as to avoid over burdening, leading to a possible loss of interest, and the most undesirable change of behavior towards memorizing and 'spotting' of questions. Perhaps one should also ask the question as to whether there is any clinical situation when all disciplines need to be integrated. Therefore, it's best to FACILITATE INTEGRATION, AND NOT FORCE IT. It can be done comfortably through case studies where problems can be analyzed with the help of other related disciplines.

The Role of an Anatomy Museum

Where dissection/prosection is an integral part of the teaching program, a museum should have a clear and specific role especially where space is a premium and research labs are tough adversaries.

The anatomy museum has to play a meaningful and complementary role to other existing facilities [11, 12]. Therefore it cannot be a simple repository for spare specimens. Ways to be unique and relevant include:

1) Having state of the art dissections for reference.
2) Presenting exposures that are difficult to achieve in practical classes, e.g. the middle ear, intracranial details, etc.

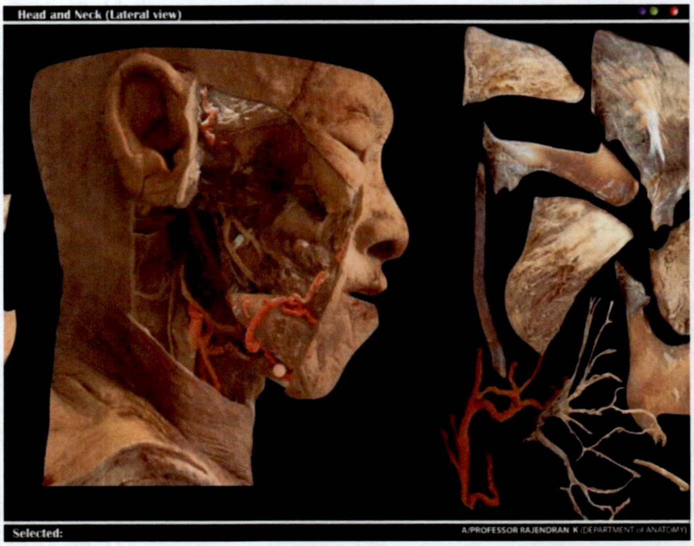

Anatomical museums complement other modes of learning with unique opportunities to access state of the art specimens with labels, and interactive "virtual dissection" computer programs.

3) Providing unique interactivity with displayed specimens on a computer to facilitate understanding beyond what is displayed. In other words, do a virtual dissection to achieve the exposure that is on display.
4) Ready access to labeled images of displayed specimens with provisions for self-testing.
5) Other multimedia material to complement the learning experience by providing basic principles and virtual anatomy 'dissection' exercises.

CONCLUSION

A discipline that is central to the medical sciences may be taught to students offering a range of courses. The visual, functional and clinically relevant nature of gross anatomy however remains relevant through it all. To achieve successful teaching of the subject it needs to be rationalized and conceived for the level (of difficulty) at which it is to be offered to different students, and be appropriately delivered to realize the intended goal of having given foundational knowledge along with approaches to appreciating, enjoying and furthering knowledge of the anatomy as well as related subjects.

REFERENCES

[1] Helen L. Naug,* Natalie J. Colson, Daniel G. Donner, Promoting Metacognition in First Year Anatomy Laboratories Using Plasticine Modeling and Drawing Activities: A Pilot Study of the ''Blank Page'' Technique. *Anat. Sci. Educ.*, 4 (2011): 231–234.

[2] Priti Pandey & Craig Zimitat, Medical students' learning of anatomy: memorization, understanding and visualization. *Medical Education*, 41 (2007): 7–14. Doi: 10.1111/j.1365-2929.2006.02643.

[3] D J A Heyling, Anatomy 1999–2000: the curriculum, who teaches it and how? *Medical Education*, 36 (2002): 702–710.

[4] HAROLD ELLIS*, Teaching in the Dissecting Room. *Clinical Anatomy*, 14 (2001): 149–151.

[5] Eun-Kyung Chung,{ Kwang-Il Nam,{ Sun-A Oh, Eui-Ryoung Han,* Young-Jong Woo, and Maurice A. Hitchcock, Advance Organizers in a Gross Anatomy Dissection Course and Their Effects on Academic Achievement. *Clin. Anat.*, (2012): 000–000. VVC 2012 Wiley Periodicals, Inc.

[6] Jun Wang, Weiguang Zhang* Lihua Qin, Jing Zhao, Shuyong Zhang, Jin Gu, Changman Zhou, Problem-Based Learning in Regional Anatomy Education at Peking University. *Anat. Sci. Educ.*, 3 (2010):121–126.

[7] Karen R. Marsh, Bruce F. Giffin,2 Donald J. Lowrie Jr.*, Medical Student Retention of Embryonic Development: Impact of the Dimensions Added by Multimedia Tutorial. *Anat. Sci. Ed.*, 1 (2008): 252–257. © 2008 American Association of Anatomists.

[8] Rajendran*, K, C K Tan and F C T Voon, "Computer graphics in the teaching and learning of anatomy", *Journal of Audiovisual Media in Medicine*, 13, no. 2 (1990): 49-52. (United Kingdom). 11. Publication no. 0028108.

[9] Yip*, G W C and K Rajendran*, "SnapAnatomy, a computer-based interactive tool for independent learning of human anatomy", *Journal of Audiovisual Media in Medicine*, 31 (2008): 46-50. (United Kingdom). (PMID: 18802832). 8. Publication no. 0370695.

[10] Rajendran, K, "Model based transparencies for gross anatomical teaching". *Journal of Audiovisual Media in Medicine,* 8 (1985): 124-126. (United Kingdom). 13. Publication no. 0028106.

[11] Pallab K. Ganguly,* Manoj Chakravarty, Nasir A. Latif, Mirghani Osman, and Marwan abu-Hijleh. Teaching of Anatomy in a Problem-Based Curriculum at the Arabian Gulf University: The New Face of the Museum, *Clinical Anatomy* 16 (2003): 256–261.

[12] Yehia M. A-H. Marreez,* Luuk N. A. Willems, Michael R. Wells. The Role of Medical Museums in Contemporary Medical Education. *Anat. Sci. Educ.,* 3 (2010): 249–253. © 2010 American Association of Anatomists.

In: Education in Anatomical Sciences
Editor: Paul Ganguly
ISBN: 978-1-62618-488-6

Chapter 8

Cell Biology/Histology Can Be the Facilitator for Integrating Biomedical Sciences in the Medical Curriculum

Bernhard H. J. Juurlink*

Professor emeritus, College of Medicine, University of Saskatchewan, Saskatoon, Saskatchewan, Canada

Visiting Professor and Founding Faculty Member, College of Medicine, Alfaisal University, Riyadh, Kingdom of Saudi Arabia

Abstract

With the continual exponential expansion of biomedical sciences information it can be extremely difficult for students to integrate the essential biomedical knowledge necessary for sound clinical practice. We already know that cells provide the functional basis of the body. We also know that the cells provide these functions in the context of other cells by being arranged in tissues and tissues into organs. It is not difficult to envision how integrative cell biology/histology might be. In this chapter I put forth the thesis that cell biology/histology is the ideal subject to integrate the biomedical sciences and give a few examples, taken form the renal and gastro-intestinal systems, illustrating how this can be done.

I have difficulty seeing boundaries between the various disciplines. This definitely is of advantage from a teaching perspective, as long as one can facilitate student learning in digestible bites. I believe that to be a good teacher one must have a sound research background, although being an excellent researcher does not necessarily result in being an excellent teacher. The reason I think that to be an excellent teacher one needs a research background is that research gives one the realization that our body of knowledge is simply the best explanation at the moment on how the world is organized - in the future, we may well have a better understanding. Without a research background we are like the water strider (Figure 1) convinced of the solidity of the foundation upon which we rest. Without a research

* E-mail address: bjuurlink@gmail.com.

background, areas of learning will tend to be treated as being fixed, much like medieval Europeans viewed Galen's body of medical knowledge. In areas such as histology there will be an overemphasis on all the minutia of terminology and concentration on structure rather than on function.

Figure 1. The water strider is completely oblivious to the elasticity of the foundation of his world. University teachers without a solid research background are in danger of being similarly oblivious of the foundational elasticity of their knowledge base.

The modern medical curriculum is faced with a number of questions. In light of the increasing body of knowledge in areas outside of biomedical sciences, one of the questions is how much basic biomedical sciences is required [1]. Some, like a former dean of mine, claim that bedside manners (i.e., the sociological/psychological aspects of medicine) are more important than biomedical sciences. To which I replied that bedside manners failed to diagnose my mother's femoral hernia despite her complaints of groin pain - the diagnosis of femoral hernia requires some knowledge of anatomy and how multiple pregnancies may affect the structural integrity of the lower abdominal wall. It is clear is that there is a strong biomedical sciences basis to health and disease although one's mental health/psychology can influence biomedical processes and, thus, health. A good physician must have a clear understanding of biomedical processes as well as how our mind may influence such processes. Indeed, a sound knowledge of biomedical sciences is especially necessary for diagnosis of difficult clinical problems [2], and as well of simpler clinical problems such as a femoral hernia.

Another question to be faced in the presence of an exponential increase in biomedical knowledge is what is the minimal biomedical science knowledge required to provide the basis of sound clinical practice. Certainly, all the details known about biomedical sciences cannot be learned (memorized) and one should focus on principles and from these principles develop a knowledge framework. This knowledge framework will allow the physician to know how to and where to access details necessary for diagnosis. Cell biology/histology is an area that can greatly facilitate students developing conceptual frameworks that enable recognition of connected patterns in health and disease.

A major problem is how to integrate the divers concepts necessary for a successful medical practice where many of the health problems arise from the modern lifestyle and affect multiple body systems rather than one specific body system.

The body does not operate as a series of semi-independent systems: anatomy (including the subdivisions of gross anatomy, embryology, neuroanatomy and histology), physiology, microbiology, immunology, etc., yet the way the human brain is organized necessitates that one must learn in digestible bites forcing considerable reductionism in approaches to understanding the functioning of the body. The development of a conceptual framework necessitates integration. In many senses it is far more important for the family physician to be able to integrate information from multiple disciplines than it is for the specialist, although the specialist without broad cross-disciplinary knowledge can be a dangerous individual.

So how does one facilitate integration of knowledge? One approach to a more integrated approach to understanding the function and dysfunction of the body is through problem-based learning (PBL) that currently is sweeping out the more traditional didactic methods. Successful PBL sessions are very tutor-dependent and can be problematical since there is a tendency for tutors to offer too much guidance or, on the other hand, to allow the discussions to go off-track.

Another pitfall in complete reliance on a PBL-based curriculum is that significant gaps may appear in basic biomedical sciences knowledge [3] - such gaps may be filled with more traditional didactic lectures [3]. The ideal curriculum, perhaps, is a combination of PBL and didactic approaches, especially in curricula developed for students that enter medicine straight from high school, as they do at Alfaisal University. One can structure the more traditional didactic session to facilitate student problem-solving in the classroom setting through posing specific questions in a carefully-sequenced manner. I have found that cell biology/histology is ideal for developing biomedical analytical skills and for the integration of physiology, cell biology, anatomy, molecular medicine, pharmacology and many aspects of microbiology/immunology if one takes a very functional approach in one's teaching. People go to family physicians or become directed to specialists because there is a functional problem with their bodies. Functional approaches should be the driving force in biomedical education.

It is my thesis that histology should be taught from a functional perspective and with this perspective one can use histology to integrate most of the biomedical sciences. Some histology texts, such as by Kiersenbaum and Tres [4], use such an approach to a certain extent, particularly in integrating histology and pathology, but much more can be done. In the pages that follow I will give a few examples illustrating the integrative aspects fostered by a functional approach to histology.

My main focus will be on renal histology as the exemplar (Figure 1), but will also give a few examples involving the gastro-intestinal tract. The kidney should be dealt with after the cardiovascular system and after the role of renin-angiotensin-aldosterone system in blood pressure regulation has already been considered. The students should also have had an introduction to signalling pathways such as the G-protein coupled pathways and the phospholipase A_2/cyclo-oxygenase (COX) pathways before studying renal function.

The Kidney as Exemplar

By this time in their studies students should have an appreciation that an essential function of an epithelium is to provide a boundary through which movements of particular substances are controlled. They will know that there will be differences in function and molecular composition between the apical and basal regions of the plasmalemma, with the differences being dependent upon the function of the epithelium. The tightness of the junctions between cells determines the extent of paracellular movement of ions and associated water relative to osmotic, ionic and voltage differences across the epithelium.

Where relatively large amounts of smaller molecules and water are required to pass, for example, the visceral epithelium of the glomerulus and associated endothelium, the involved epithelia should offer little resistance to movement; thus, in the renal corpuscle, there is the presence of podocytes and associated pedicles forming filtration slits and fenestration of the endothelium.

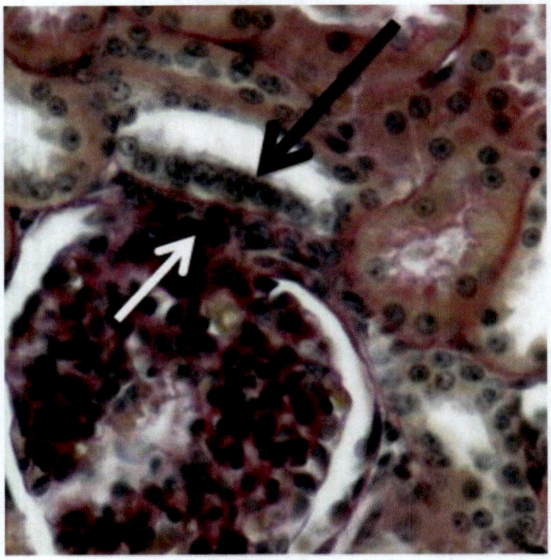

Figure 2. A section through the kidney that shows a renal corpuscle, a juxtaglomerular apparatus and portions of the proximal and distal convoluted tubules.The large black arrow indicates the macula densa while the smaller white arrow indicates the juxtaglomerular cells. To the lower right of the macula densa is a section through a proximal convoluted tubule and below this, next to the renal corpuscle, is a section through a distal convoluted tubule. Section treated with periodic acid-Schiff and hematoxylin.

Students should appreciate that the negative charges associated with the filtration slits and within the common basal lamina interferes with movements of most proteins due to proteins' overall negative charges that, amongst other things, prevents clogging up of the filtration device. These are nice examples of how the specific molecular arrangements at specific sites influence the function of the body as a whole. Students should begin to reflect on what may happen to the kidney's filtration function if such negative charges disappear. Further, they should appreciate that the ability to differentially control the caliber of the afferent and efferent arterioles results in an ability to manipulate the glomerular filtration pressure.

Proximal convoluted tubule: One can teach much of histology through asking appropriate questions. Upon learning that it is in the proximal convoluted tubule where most of the salts, amino acids and glucose present in the filtrate are reabsorbed, students should immediately begin to think about uptake mechanisms, especially in the context of Na^+-dependent transport mechanisms and specific ion transporters. To pack high numbers of uptake mechanisms requires increase in surface area and students should be able to deduce that the apical epithelium is characterized by being comprised of microvilli. Students should know that: i) Na^+-dependent uptake increases intracellular Na^+ concentration, and ii) much of the Na^+ present in the filtrate should re-enter body fluids. Hence, they should deduce that the majority of Na/K pumps are located in the basolateral parts of the epithelial cells and that these provide the motif force for ion transport from (and to) the filtrate. They should also be able to deduce on first principles that all this Na/K pump activity requires considerable ATP and that one would expect large numbers of mitochondria near the vicinity of the pumps. One would also expect a large surface area for the basolateral membranes to allow incorporation of Na/K pumps and efflux mechanisms for such molecules as glucose and amino acids. Since the basal parts of the epithelium needs to be anchored to a basal lamina it should not come as a surprise to the student that membrane surface area is increased by membrane in foldings rather than through the formation of microvilli. Students should also be able to deduce on first principles that the apical glucose uptake transporter is Na-dependent since glucose is transported mainly against its concentration gradient, whereas, the basolateral transporters should be Na-independent allowing energy-free movement down the glucose concentration gradient. Thus, students should have an appreciation that function is what drives structure both at the macromolecular and the molecular levels.

Not only is the function of the proximal convoluted tubule energy intensive but upon learning about the function of ion uptake, particularly Na^+ by the distal convoluted tubule and the collecting duct system, students can deduce that in the resting body the kidney, on a per gram tissue basis, has a very high rate of energy utilization; hence, students should not be surprised that this exceeds even the brain's energy utilization but that it is lower than that of the heart's energy utilization.

The Juxtaglomerular Apparatus. One should go beyond simply mentioning that the macula densa cells detect urinary flow and, under conditions of low flow, signal to the juxtaglomerular smooth muscle cells to release the protease renin. The juxtaglomerular apparatus offers a major opportunity to integrate many of the aspects of blood pressure control. The discussion below deals only with the major pathways that are known and the effects of acute changes, but not chronic changes that form adaptations to an altered homeostasis. The latter can be more appropriately dealt with during later aspects of the medical curriculum.

Once students have covered the cardiovascular physiology they should know that the kidney, by secreting renin, plays a major role in the regulation of blood pressure. Hence, they should have some familiarity with the renin-angiotensin-aldosterone system and the use of the angiotensin converting enzyme (ACE) inhibitors as a treatment for hypertension. One can test the students through appropriate questions to remind them of this aspect of blood pressure regulation (note Figure 3). No doubt, while discussing one of their PBL modules covering hypertension they have encountered β-adrenergic receptor blockers as another possible treatment for hypertension. This latter may be somewhat puzzling for them since they probably are likely familiar with the fact that stimulation of β-adrenergic receptors in arterial smooth muscle causes vasorelaxation because of PKA-dependent phosphorylation of myosin light chain kinase (MLCK) that inhibits MLCK's activity. Thus, students integrating their cardiovascular physiology and pharmacology become confused over why blocking β-adrenergic receptor should result in lowering of blood pressure rather than raising it: the key, of course, is in understanding the regulation of renin secretion.

One should start this discussion of the juxtaglomerular apparatus by reminding students that juxtaglomerular cells, present in the afferent arteriole at the glomerular pole, synthesize renin and that renin secretion is governed by a cAMP/Protein Kinase A (PKA) pathway.

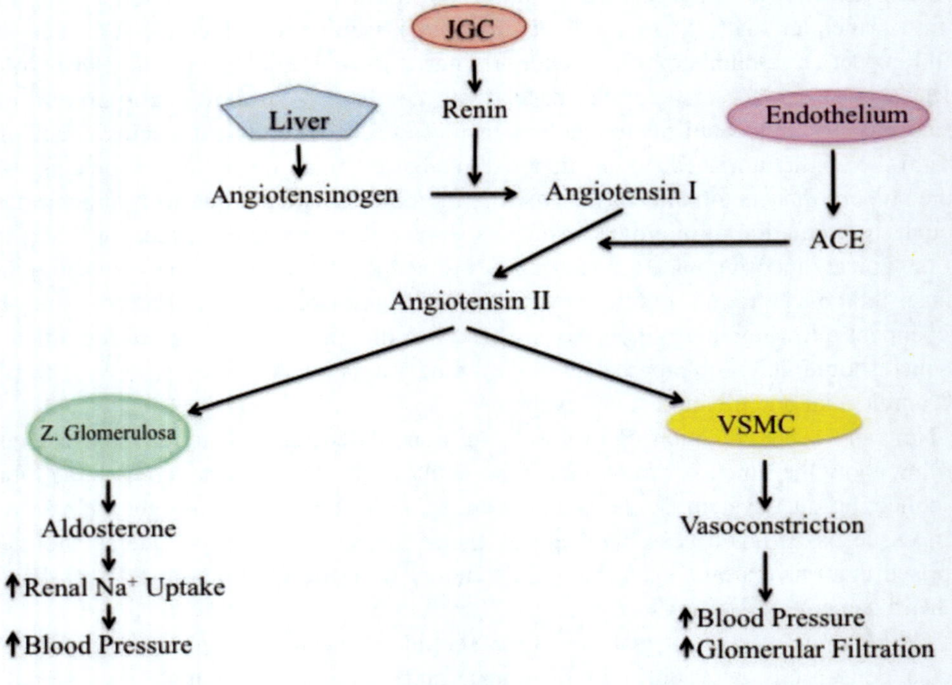

Figure 3. Schematic of the renin-angiotensin-aldosterone system and tissues involved in blood pressure regulation that illustrates what students should be reviewing. The juxtaglomerular cells (JGC) in response to signals secrete renin that cleaves a peptide off angiotensinogen, produced by the liver, to form angiotensin I. Angiotensin I is then cleaved by angiotensin converting enzyme (ACE) produced by vascular endothelial cells (predominately pulmonary) to form the active hormone angiotensin II. Angiotensin II then acts on AT1 receptors to activate the Gq component of the G-protein coupled pathway in adrenal zonaglomerulosa cells and in vascular smooth muscle cells causing activation of phospholipase C resulting in the release of inositol IP_3. IP_3 then acts on receptors on the smooth endoplasmic membrane causing Ca^{2+} release. The rise in intracellular Ca^{2+} causes contraction of smooth muscle cells resulting in an increase in blood pressure as well as an increase in glomerular filtration because of greater contraction of the smooth muscle in the efferent arteriole compared to the afferent arteriole supplying the glomerulus. The rise in intracellular Ca^{2+} in glomerulosa cells results in the synthesis and release of aldosterone that then acts upon receptors in the collecting tubular system as well as the distal convoluted tubule to increase Na/K pump activity, thereby increasing Na^+ uptake (and K^+ excretion) resulting in increased plasma volume and, thus, increased blood pressure.

Since many of the details of this regulation are not completely understood, it is not necessary to go beyond the fact that rises in cAMP, due to activation of Gs pathway, causes PKA to phosphorylate downstream proteins that result in renin secretion. Students should already be familiar with a broad outline of the pathway where renin cleaves angiotensinogen to angiotensin I that, in turn, is cleaved by angiotensin converting enzyme (ACE) to the angiotensin II that activates angiotensin I (AT1) receptors on the vascular smooth muscle and adrenal cortical zona glomerulosa cell (Figure 3). Activation of AT1 receptors in both the zonaglomerulosa and vascular smooth muscle cells activates the Gq path of the G-protein coupled signalling resulting in an increase in intracellular Ca^{2+}. Such reminders need not necessarily be in the classroom but could form a component of the PBL problem with which they are currently dealing. Students should be reminded that increased intracellular Ca^{2+} causes synthesis and release of aldosterone by the adrenals and contraction of vascular smooth muscle cells (more in the efferent arteriole than the afferent arteriole) resulting in a general increase in blood pressure and increased glomerular filtration. This is a review in the context of the histological arrangements of the involved tissues and a brief reminder of this pathway will reinforce information they learned in other contexts. At some point one should also remind the students of what they have learned in physiology that aldosterone acts on the principle cells of the collecting ducts and on the distal convoluted tubular cells to increase the Na/K pump activity in the basolateral membranes and to increase the epithelial Na channels in the apical cell membranes of these cells, i.e., aldosterone causes increased Na^+ uptake plus associated water and, thus, increases plasma volume and, thereby, blood pressure.

Students should then be introduced to the concepts that the juxtaglomerular cell cAMP/PKA pathway can be activated by signals coming from either the sympathetic nerve fibres or from the macula densa cells (Figure 4). Students should already be familiar with the flight/fright/fight/frolic aspect of the sympathetic nervous system and should already be familiar with catecholamines activating α-adrenergic receptors in most vascular smooth muscle cells resulting in an increase of intracellular Ca^{2+} via the Gq/IP_3 G-protein coupled pathway and that increased intracellular Ca^{2+} initiates the contraction of vascular smooth muscles thereby increasing blood pressure.

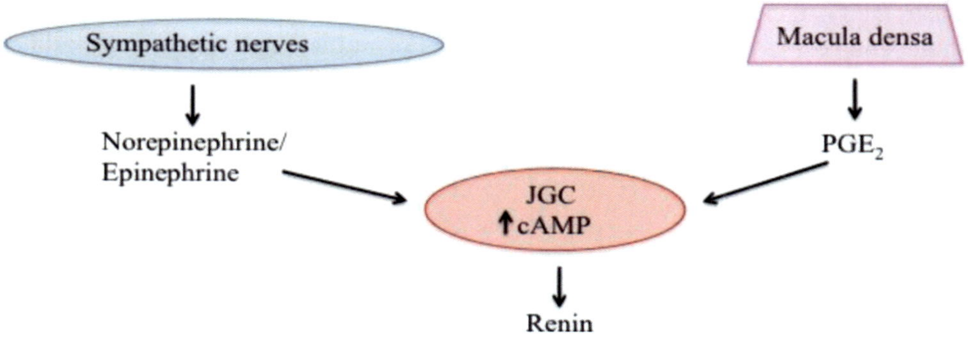

Figure 4. Schematic that illustrates the dual control by sympathetic nerves and the macula densa of renin secretion by the juxtaglomerular cells (JGC). Both catecholamines and PGE_2 activate G-protein coupled receptors that activate Gs, thereby increasing cAMP levels that,by activating PKA, initiates the pathway that leads to renin secretion.

The students now know that under flight/fright/fight/frolic conditions there is an increase in blood pressure due to α-adrenergic receptor activation and they also know that angiotensin II will cause an increase in blood pressure. Therefore, it should not come as a surprise that sympathetic nerve activity will activate β-adrenergic receptors on juxtaglomerular cells resulting in increased renin secretion (Figure 4).

This is in keeping with the flight/fright/fight/frolic perspective of the function of the sympathetic arm of the autonomic nervous system. Knowing that β-adrenergic receptors are involved in one pathway regulating renin secretion, thus, eliminates the students' puzzlement to the blood pressure lowering effects of β-blockers.

This is perhaps is a good time to remind the students that β-adrenergic receptors coupled to Gs are present in the coronary arteries as well as arteries supplying skeletal muscles. This should come as no surprise since under conditions requiring sympathetic activation we wish to ensure adequate blood flow to the heart and skeletal muscles.

One could also remind students that cardiac myocytes also express β-adrenergic receptors linked to Gs: here, increased PKA activity, amongst other things, results in phosphorylation of plasmalemmal L-type channels as well as the sarcoplasmic ryanodine channels promoting increased cytosolic Ca^{2+} influx in response to membrane depolarization - this increased intracellular Ca^{2+} results in stronger contractions. Thus, β-blockers decrease renin secretion and reduce cardiac contractility, thereby decreasing blood pressure (and also reduce cardiac blood supply). Understanding the multiple sites of action of β-blockers will better prepare the students to later understand some of the possible adverse effects of β-blockers.

One could then deal with the signal given by the macula densa cells to promote renin secretion by the juxtaglomerular cells. This signal is prostaglandin E2 (PGE_2) produced by a pathway that requires cyclo-oxygenase-2 (COX2) activity. For recent reviews see [5, 6]. PGE_2 diffuses to the juxtaglomerular cells where it activates the E-prostanoid receptor-2 (EP_2). EP_2 is linked to the Gs component of the G-coupled protein pathway; hence, PGE_2 promotes activation of the cAMP/PKA pathway in the juxtaglomerular cells - the effect is the same as what is seen with β-receptor activation, i.e., increased renin secretion. There is an elegant symmetry to this dual regulation of renin secretion that should be appreciated by the students (Figure 4). Parenthetically, students will also, at some point, learn that PGE_2 will stimulate the Gs G-protein coupled signalling in stomach epithelial cells to promote mucous synthesis and secretion, and if this was covered prior to the renal system one could remind students of this, reinforcing the notion that there may be some commonalities in regulating protein secretion.

Ideally the students should have had an introduction to inflammation including the central role of nuclear factor kappa B (NFκB) in promoting pro-inflammatory gene expression with NFκB activation promoting the expression of cyclo-oxygenase-2 (COX2) – molecular medicine is an ideal area for this. Students likely would have encountered the idea that eicosanoids produced by the cascade initiated by COX2 on arachidonic acid generally promote inflammation and that non-steroidal anti-inflammatory drugs act by inhibiting COX enzymatic activity. However, one could remind the students that there are non-inflammatory roles for eicosanoids as evidenced in the promotion of renin and mucous secretion. The students should already have encountered that rises in intracellular Ca^{2+} can lead to activation of phospholipase A2 (PLA_2) whose action is to release the fatty acid present at the SN-2 position of membrane phospholipids. The more common fatty acid at the SN-2 position is the omega-6 polyunsaturated (PUFA) acid arachidonic acid, but it can also be occupied by omega-3 PUFAs such as eicosapentaenoic acid (EPA) or docosahexaenoicacid (DHA).

The students should have already encountered the fact that the COX constitutively present in many tissues is COX1 and that COX2 is induced early in the inflammatory process. It may come as a surprise to students that COX2 is constitutively present in a number of cell types, including the macula densa cells. This information on the constitutive presence of COX2 in a number of different tissues should cause students to think about possible side-effects with selective COX2 inhibitors (they should reflect on renin secretion, for example) and begin to wonder whether the use of nonsteroidal anti-inflammatory drugs for a condition such as arthritis may influence gastric and renal function. These reflections emphasize that one is treating the body as a whole, not just some component of body function/dysfunction and, again, that the actions of drugs may be complex.

Somewhere in their learning students ought to have encountered the information that whether arachidonic acid or the omega-3 PUFAs occupy the SN-2 position of membrane phospholipids is dependent on the relative dietary intake of omega-3 and omega-6 fatty acids: this is one way that diet can ameliorate inflammation since products of arachidonic acid tend to be inflammatory while products of EPA and DHA tend to be anti-inflammatory. Furthermore, omega-3 PUFA dietary deficiency is correlated with hypertension and supplementation of omega-3 PUFAs results in better blood pressures. Students should begin to reflect upon possible ways this may happen and reflect on how diet may influence many aspects of our body's functions.

This is probably an appropriate time to consider what the macula densa cells sense and how this information is transduced to activation of the pathway that leads to PGE_2 synthesis. What is essentially sensed by the macula densa is the concentration of NaCl (in particular, it appears the concentration of Cl^-). Although it is a little more complicated than this since the cells also detect the rate of fluid flow through its primary cilium - but this can be ignored since it has, at the moment, no relevance to therapeutic interventions. Decreased tubular NaCl concentrations results in cell volume shrinkage that is accompanied by increased intracellular Ca^{2+} [7]. On the other hand, increased tubular NaCl concentrations have been shown to lead to macula densa cell swelling but no change in intracellular Ca^{2+}. Students from this information should begin to think of the possibility that cell shrinkage opens Ca^{2+} channels. Students are already familiar with ligand-gated ion channel as well as voltage-gated ion channels and should suspect that channels can also be mechanically gated. One can confirm their suspicions that there are mechanically-gated Ca^{2+} channels in the plasmalemma of macula densa cells. Thus, this information increases their repertoire of knowledge on ion channel gating mechanisms. The rise in intracellular Ca^{2+} is followed by increased PGE_2 synthesis. Upon questioning what the relationship between a rise in intracellular Ca^{2+} and increased synthesis of PGE_2 may be, the students should be reminded that macula densa cells express COX2. This should cause students to consider that the rise in intracellular Ca^{2+} activates phospholipase A_2, thereby releasing arachidonic acid, a substrate for COX enzyme activity - this ultimately leads to synthesis of PGE_2.

At this time, students should be thinking about how ion movements will lead to changes in cell volume. They should immediately start reflecting on how ions can move from the tubular filtrate to the interstitial space. They should immediately consider that the driving force would be Na/K pumps localized to the basolateral membranes of the macula densa cells. They may reflect on what may happen if there is a constant efflux of ions and associated water from the basolateral cell membranes but the counter-balancing influx of ions and associated water through the apical cell membranes drops due to lowered tubular NaCl levels. As a background the students should be informed that the macula densa cells, unlike most, have little ability for volume regulation. They should be reflecting on what happens if one removes more osmolytes from a cell than enter the cell – if they have learned their basic principles well the answer they will come up with is cell shrinkage. Whether this is the actual mechanism of shrinkage is still to be determined, but this exercise reinforces the development of a mechanistic perspective.

From physiology the students will know that furosemide blocks the actions of the Na-K-2Cl symporter in the thick ascending loop of Henle and, thus, decreases the osmotic gradient in the medulla resulting in more fluid loss that should decrease tissue edema and also, at least in principle, significantly decrease blood pressure due to lowered plasma volume. What generally is not covered is that the Na-K-2Cl symporter is also present in the apical membranes of the macula densa cells and most of the Na^+ uptake is via this symporter. Blocking the principal component of the NaCl uptake mechanism should greatly interfere with tubuloglomerular feedback and students might speculate that renin secretion should be increased and, indeed, this is the case. This explains why furosemide treatment can have significant effects on edema but relatively little effect on blood pressure.

Of course, anyone that delves into the research literature realizes that the tubuloglomerular feedback is more complicated than the discussion above and what is ignored includes the role of nitric oxide, angiotensin II feedback, etc. This more complicated look at the relationship between the kidneys and the cardiovascular system is best left to later in the curriculum for the students that intend to become specialists in the renal and/or cardiovascular systems.

The approaches illustrated above greatly assist students in developing a knowledge framework in a very integrated fashion. One might think that the kidney is more ideally suited for such an approach than most other systems. This is not the case. I will briefly give examples of the gastro-intestinal system.

Reflections on the Role of Mucins in the Gastro-Intestinal System (GIS)

Students should have already considered the properties of mucins in molecular medicine and how such highly glycosylated proteins order water, resulting in the formation of a very viscous fluid with only a portion of the water being free where it forms channels that allow movement essentially only of small molecules. With this background, after considering digestion of macromolecules into molecules and how these molecules must diffuse through the very viscous mucous covering the enterocytes, one can question the students on what may facilitate movement of small molecules. Encountering the muscularis mucosa should enable students to come up with the idea that muscularis-driven movement of villi causes stirring of the viscous fluid in the vicinity of the apical portions of the enterocytes, thus facilitating movement of small molecules enhancing the probability of the small molecules making contact with the transport mechanisms present in the microvilli. After some probing students should also come up with the idea that villus movement also facilitates movement of the fluids within the lacteals present within the lamina propria of the villi, thereby facilitating lacteal emptying into lymphatic venules.

One can also discuss how the presence of a viscous mucous covering decreases the probability of intestinal bacteria making contact with the epithelium thereby decreasing the likelihood of gut bacteria entering our tissues. In the oral cavity and esophagus one could discuss how having a stratified squamous epithelium covered with mucous offers protection against mechanical damage. The stratification of the epithelium decreases the likelihood of the epithelial barrier being penetrated food and the mucous reduces the resistance of the epithelium to foods on their passage to the stomach. After discussing the necessity of a low pH in the stomach for digestive purposes including denaturation of proteins making them more susceptible to proteases and the requirement for an acid environment to allow pepsin activity that is necessary for initiating collagen digestion, one should raise the question on how the epithelium protects itself from acid attack. This is an opportune time to discuss how mucin composition may vary according to the specific mucosal surface: that specific mucins present along the stomach epithelium together with bicarbonate diffusing from the epithelium allows the neutralization of the hydrochloric acid as it diffuses into the mucous. One can then bring up the fact that in many people *Helicobacter pylori* reside in the stomach and query students on how it is possible that a bacterium can survive in such a harsh environment. They should be able to come up with the idea that the bacterium is residing in the mucous where it is being buffered by some alkaline substance. A Google search by the students should lead them to the information that the bacteria break down urea to form ammonia; hence, the bacteria are nestled in a viscous mucous and secrete ammonia to counteract the hydrochloric acid diffusing into the mucous. This reinforces several concepts they have already encountered elsewhere in physiology and molecular medicine including that urea at neutral pH can pass down the concentration gradient through cell membranes and that urea is essentially a storage form of ammonia.

Students can also be asked to speculate why Barrett's esophagus where, although the epithelium has undergone metaplasia to form gastric epithelium, is still susceptible to acid reflux with the consequence that there is an increased probability of esophageal cancer formation. One would expect students to answer that possibly the appropriate mucins are not being expressed by the epithelium or possibly inadequate bicarbonate secretion by the epithelium. A search of the literature suggests that the reason for increased probability for cancer formation is not known, although there are hints that it may be due to the proximal segments of Barrett's esophagus expressing intestinal rather than gastric mucins- intestinal mucins do not offer good protection against acidity. Nevertheless, what is important is that students when encountering new knowledge should attempt to see how it may make sense with what they already know, i.e., there should be an active constructive modification of their knowledge framework.

CONCLUSION

Cells mediate the functions of the body. It is the molecular arrangements of the cells, their metabolism and their secretory products as well as the relationships of the cells to one another and to their secretory products that enable these functions. Cells have these functions by being arranged into tissues. The tissues form the organs. *Apriori*, one can readily see how histology can integrate the biomedical sciences. I trust that the few examples given will cause the reader to consider that histology is an area that is conducive to integration of biomedical sciences.

REFERENCES

[1] Finnerty EP, Chauvin S, Bonaminio G, Andrews M, Carroll RG, Pangaro LN. Flexner revisited: the role and value of the basic sciences in medical education. *Acad. Med.*, 2010; 85: 349-355.

[2] Woods NN, Brooks LR, Norman GR. The role of biomedical knowledge in diagnosis of difficult clinical cases. *Adv. Health Sci. Educ. Theory Pract.,* 2007; 12: 417-426.

[3] Smith HC. A course director's perspectives on problem-based learning curricula in biochemistry. *Acad. Med.*, 2002; 77: 1189-1198.

[4] Keirzenbaum AL, Tres LL. Histology and Cell Biology. An Introduction to Pathology. Third Edition, Philadelphia, PA, USA, Elsevier Saunders. 2011: 720.

[5] Peti-Peterdi J, Harris RC. Macula densa sensing and signaling mechanisms of renin release. *J. Am. Soc. Nephrol.*, 2010; 21: 1093-1096.

[6] Kim SM, Briggs JP, Schnermann J. Convergence of major physiological stimuli for renin release on the Gs-alpha/cyclic adenosine monophosphate signaling pathway. *Clin. Exp. Nephrol.*, 2012; 16: 17-24.

[7] Liu R, Persson AE. Simultaneous changes of cell volume and cytosolic calcium concentration in macula densa cells caused by alterations of luminal NaCl concentration. *J. Physiol.*, 2005; 563: 895-901.

In: Education in Anatomical Sciences
Editor: Paul Ganguly

ISBN: 978-1-62618-488-6

Chapter 9

ANATOMICAL EDUCATION - EMBRYOLOGY: INTEGRATING CLINICALLY ORIENTED EMBRYOLOGY INTO AN ORGAN SYSTEMS-BASED CURRICULUM

Michael J. Rindler*
Department of Cell Biology, New York University School of Medicine, New York, US

ABSTRACT

Embryology has traditionally been taught as an adjunct of Gross Anatomy, emphasizing visualization of developmental processes to assist students in their understanding of anatomical relationships. However, other major components of Embryology courses, such as molecular mechanisms of development and congenital malformations, are not a good fit with Gross Anatomy. The process of curriculum reform at New York University School of Medicine, where the basic science curriculum was reorganized into an organ systems-based approach, was accompanied by the restructuring of the Gross Anatomy and Embryology course material. As part of this process, much of Embryology was repositioned to be taught in conjunction with organ system physiology, histology, pathology, and pathophysiology. The reorganization allowed for clinically oriented embryology to be taught at a more sophisticated level. However, the curriculum reform was also accompanied by a reduction in basic science teaching time. Although the process eliminated some of the redundancies in the curriculum, it has proven challenging to maintain a comprehensive Embryology program. Moreover, assessment of students on examinations has become more problematic as Embryology is now often a minor segment of the organ-based modules. Embryology material is also revisited in the clinical clerkships (obstetrics, pediatrics) and electives, helping to accomplish the goal of integrating basic science throughout the four year medical school curriculum.

* E-mail address: Michael.Rindler@nyumc.org.

Introduction

Major changes in medical education in the United States have taken place over the past decades. In recent years, there has been an increased emphasis on clinical core competencies, as defined by the Accreditation Council for Graduate Medical Education (ACGME; http://www.acgme.org/acgmeweb). In practice, many institutions have developed their own set of expectations for undergraduate medical education that parallel those of the ACGME in the areas of medical knowledge, patient care, practice based learning, interpersonal and communication skills, professionalism, and systems based practice. The focus on core competencies has meant that basic science courses have had to become more clinically oriented and case- or problem-based than was generally the standard in traditional curricula. In addition, a number of institutions, including New York University School of Medicine (NYUSOM), have reduced the time allotted for the basic science curriculum overall. The decreased time (in our case from 2 to 1.5 years) has forced deans of medical education and course directors to make difficult choices as to what will be taught and to define what the objectives and core subject matter should be in each discipline. The adoption of NYUSOM's Curriculum for the 21st Century (C21) in 2010 altered the way we teach and think about Embryology in the basic science program. In planning the changes in the curriculum, the issue was raised as to whether Embryology should be included. After all, it is taught very superficially or not at all in some institutions. The answer to this question is that Embryology is a rapidly advancing field, with breakthroughs in molecular genetics, prenatal testing, and neonatal surgery, as well as embryonic stem cell technology, occurring all the time. Indeed, the Nobel prizes for Medicine in 2012 were awarded to John Gurdon and Shinya Yamanaka, pioneers of stem cell biology, and that in 2010 to Robert Edwards for the development of human *in vitro* fertilization. As discussed a decade ago by Carlson in his prescient article [1], Embryology (often called Developmental Anatomy) is a poster child for a 21st century discipline that should remain as part of the core curriculum for medical students. Embryology not only forms the basis for understanding obstetrics, pediatrics, and pediatric surgery, it also is very relevant to adult disease. More than 5% of us will ultimately be diagnosed with a birth defect affecting our health. Furthermore, there has been considerable progress in the care and treatment of premature infants and children with severe metabolic diseases. Incredible advances in surgical techniques have allowed the survival of infants with severe cardiac and gastrointestinal malformations, among many others. Many of these children are living to adulthood, often presenting new medical challenges for physicians. The decrease in classroom time for teaching Embryology has jeopardized our ability to give students a comprehensive course on the subject. Indeed, it is a challenge to include all of the subject subcategories listed in the guidelines of the American Association of Clinical Anatomists [2]. Moreover, the evolution of the USLME Step I examination, taken by all medical students in the U.S., towards a more integrated format has meant that a clinically oriented approach now should be the standard for Embryology courses. In this context, it is clear that the standardized tests will focus on the most common congenital malformations, as opposed to rare and more obscure defects. Therefore, it is important that students obtain a strong foundation based on sound clinical reasoning rather than an encyclopedic but superficial knowledge of the subject matter.

WHERE DOES EMBRYOLOGY BELONG IN THE CURRICULUM?

As is common in many medical schools, the Embryology course at NYUSOM had been previously taught in an integrated fashion with Gross Anatomy and was given in the first semester of medical school. Learning embryology was extremely useful to students studying anatomy. Embryology helped them to visualize organ relationships in the adult cadavers they were dissecting, as well as to rationalize innervation patterns and vascular supplies. Many of the lectures were given by Anatomy faculty who incorporated enough anatomy in their presentations to ensure relatively seamless integration. However, major segments of the Embryology course material, including principles of developmental biology and clinical applications, were difficult to fully integrate, despite the fact that the Gross Anatomy course itself also made use of clinically oriented case studies as a vital component of the didactic program. Some Embryology case studies were taught (by clinicians), but not at a very sophisticated level due to the inexperience of the first year, first semester students. This dichotomy between visually oriented and clinically oriented Embryology is well known (e.g., [1]). The C21 curriculum begins with a Core Foundations of Medicine block consisting of genetics, biochemistry, molecular and cell biology, immunology and microbiology. It also includes histology of the basic tissues. Gross Anatomy had always been taught in four separate blocks -- Trunk, Pelvis and Perineum, Head and Neck, and Limbs. The planners of the new curriculum decided initially to break up the course and have students do anatomy dissections at different times during the basic science program before the organ system modules. The dispersion of Gross Anatomy was not as successful as expected, in part because the compressed time did not afford students the opportunity to adequately review the anatomy they had learned months before. Beginning in 2012, the Trunk and Pelvis block were combined into a single four-week segment given at the end of the first semester. The Limb and Head and Neck blocks were moved and taught consecutively at the beginning of the third semester (2nd year), followed by the Neuroanatomy section of the Neuroscience module. The remainder of the new curriculum is organ system-based (Core Principles of Medicine) and includes Cardiovascular, Respiratory, Renal, Endocrine/Reproduction, Gastrointestinal, Neuroscience, Hematology/Oncology, and Musculoskeletal modules. These modules begin in the second semester and are designed to integrate the physiology, pharmacology, histology, pathology, and pathophysiology of each organ.

In considering where to teach the different segments of Embryology, a major consideration was how relevant each segment was to Gross Anatomy and to what extent embryological material was already being reviewed or taught in our previous second year organ system-based clinical curriculum. The early stages portion of the course includes fertilization, implantation, gastrulation, neurulation, embryonic folding and somite formation in a mixed lecture/small group conference format that takes advantage of computer animations and the expertise of the faculty in NYUSOM's Developmental Genetics program. It features several cases and a discussion of the principles of developmental biology and concludes with a lecture on Embryonic Stem Cells. We therefore incorporated the Early Stages content into the Core Foundations module at the end of a unit of Cell Biology that also contains lectures on cell cycle, cell death (including an example of programmed cell death during embryonic development), cell transformation, neoplasia, and signaling, as well as a case study on the use of anti-HER2 therapy in treating breast cancer. This unit already had a significant emphasis on growth factors (e.g., TGFβ, FGF, EGF and VEGF) and their signaling pathways and came after students had studied transcription factors, molecular biology and genetics. Therefore, the developmental biology component of the Early Stages of Embryology fit well here as did the topic of embryonic stem cells, which includes growth factor-induced differentiation, histone modifications and other transcriptional regulatory mechanisms.

Two parts of Embryology were kept as integrated segments of Gross Anatomy – Gastrointestinal (GI) development and Head and Neck development. The anatomical relationships of the organs of the GI tract follow directly from the embryology. Similarly, learning the cranial nerve innervation patterns as well as the muscles and bones of the head is facilitated by understanding the underlying embryology. The other segments of the course were integrated with their organ systems. These consisted of cardiovascular, respiratory, renal, and genital development. In some units, such as Cardiovascular and Reproduction, there existed lectures in the previous curriculum that contained elements of Embryology. These were combined with the former Embryology lectures and case studies. A number of lectures were considered redundant and dropped from the schedule during this process. In addition, some subject areas contained embryology material presented in various contexts. One example is a lecture and small group conference in the Cardiovascular module on heart murmurs that includes listening to murmurs, some of which were caused by congenital malformations (e.g., ventricular septal defect).

Another example is the Histology lecture and laboratory on female reproductive system (including placenta) and a lecture on placental pathology. Here placenta development, structure, function, and anomalies (e.g., hydatidiform mole, placenta previa) are presented. Students are given relevant handouts to read and digitalized Histology and Pathology slides to view in their laboratories using the Virtual Microscope program developed by the Division of Medical Informatics [3]. In addition, there is a lecture on perinatal pathology in which, among other topics, the pathogens that cross the placenta (and the consequences for the fetus), hemolytic disease of the newborn, autoimmune disease, and pre-eclampsia, are discussed. All of these topics were subjects previously covered superficially in the *Embryology* course before it was integrated during the C21 curriculum reform. These lectures, two of which are given by clinicians, integrate embryology without being designated as *Embryology* in the course schedule. In addition, many of our Histology tissue and organ-system units, which were written by our own faculty, also contain information on the embryological origins of different cell types. The goal is for students to view placental embryology, physiology, histology, and pathology/pathophysiology as an integrated, unified block.

THE EMBRYOLOGY CURRICULUM

The goals and objectives of the Embryology course and the content in each module of the curriculum are presented in Table 1. Overall our objectives are more conceptual and clinical than they are visual. We never ask students to identify structures in embryos on examinations (see Assessment below). We do, however, expect students to know and understand all of the major congenital abnormalities of the organ systems as well as the pathophysiology and often, options for treatment. For example, students are expected to learn about pulmonary development. This includes prenatal and postnatal development, lung morphology, a basic knowledge of branching morphogenesis and its control by growth factors, and surfactant production and properties.

The lecturer is a pulmonologist who presents the material as a series of cases. Students are expected to understand the causes and consequences of maldevelopment of the lungs and the trachea/esophagus as well as the standard treatments. The development of the brain, including the origin of newly discovered neural stem cells, is given as part of the Neuroanatomy segment of the Neuroscience course, which also has undergone a significant decrease in allotted hours. The lecturer spends part of an overview lecture (of the Neuroanatomy block) reviewing neural tube formation and presenting a brief overview of the embryology. Because of the compression of the Neuroscience block, developmental defects are not presented. However, students do study neural tube development and the causes of anencephaly and spina bifida as part of the Early Stages of Embryogenesis section now part of the Core Foundations of Medicine module (see above). Certain Embryology topics had to be dropped from the curriculum due to the substantial reduction of time for teaching, especially for the Head and Neck and Limb sections of Gross Anatomy. These included development of the eye and development of the limbs. Eye development was part of a series of lectures and conferences on the anatomy of the orbit given by an ophthalmologist who is also an anatomy instructor. When the teaching time for this series was reduced, he elected not to present eye development or the congenital malformations associated with it.

Table 1. Goals and Objectives of Embryology Content in the Basic Science Curriculum

The overall goal of the basic science program is to provide students with the foundation and the tools to become independent, lifelong learners who can confidently use the information to interpret and solve biomedical and clinical problems at any point in their careers.
Goals of Embryology (To be able to):
1. Describe and illustrate the development of the early embryo and the major organ systems. 2. Analyze how the anatomic relationships of the organ systems derive from their embryological origins. 3. Explain how cell fate during embryonic development is determined. 4. Discuss the major congenital malformations associated with errors in development, their pathophysiology, the underlying molecular mechanisms responsible, and available options for treatment.
Organ System Embryology Objectives
I. Early Stages of Embryogenesis Objectives:
1. Define the overall sequence of fetal development. 2. Describe the major steps in fertilization and implantation of the embryo. 3. Illustrate the early stages of embryonic development and the major defects associated with them. 4. Define the molecular mechanisms that underlie cell fate determination and embryonic tissue differentiation. 5. Explain the methodologies for generating embryonic stem cells and their potential therapeutic utility.
II. Gastrointestinal System Development Objectives:

1. Describe the development of the caudal foregut, midgut, and hindgut and the structures that derive from them during embryogenesis. 2. Analyze how the development of the gut explains adult gastrointestinal anatomical relationships, vascular supply and innervation patterns. 3. Discuss the causes and consequences of the major abnormalities associated with defective development of the gastrointestinal tract and associated organs.
III. Cardiovascular System Development Objectives:
1. Describe the development of the heart and compare the fetal and postnatal circulations. 2. Comment on the major congenital malformations associated with abnormal development of the heart as well as their underlying developmental defects. 3. Discuss the diagnosis of patients with congenital heart disease from a clinical perspective. 4. Trace the normal development of the aortic arches and their derivatives as well as the venous system in the vicinity of the heart and explain the consequences of abnormal development.
IV. Respiratory System Development Objectives:
1. Describe the stages of embryological development of the respiratory tract, lungs, and diaphragm. 2. Comment on the causes and consequences of inadequate lung or trachea development from a clinical perspective.
V. Renal and Urinary System Development Objectives:
1. Describe normal development of the kidneys and urinary bladder. 2. Discuss the congenital malformations associated with renal development and their clinical consequences.
VI. Genital System Development Objectives:
1. Describe the development of the male and female reproductive systems. 2. Discuss the causes of anomalies associated with disorders of sexual development. 3. Comment on the maternal-fetal relationship during gestation. 4. Explain the role of the placenta in fetal health and disease as well as the pathology associated with poor placental function or development. 5. Comment on the underlying principles that govern the roles of the environment and genetics in causing congenital malformations and explain the techniques used to detect malformations prenatally.
VII. Head and Neck Development Objectives:
1. Describe the development of the brain and structures in the head and neck region. 2. Discuss the major malformations associated with defective development of head and neck structures and their consequences. 3. Comment on the surgical procedures available for treatment of patients with craniofacial malformations.

Our future goal is to substitute a short animation and online module to replace the lecture. Because we have a world-renowned Reconstructive Plastic Surgery department, we retained a lecture by the department chair on head and neck abnormalities, which emphasizes surgical correction of malformations as well as the basic embryology and molecular genetics from a clinical perspective. A lecture on the development of the limbs, given by a clinician who is a researcher in the field, was also deleted from the new curriculum. This lecture had focused on the molecular pathways controlling limb development and axis formation. The limb is an excellent example of how morphogens and transcription factors specify anterior-posterior, dorsal-ventral and proximal-distal structures in the embryo. The correlation with Gross Anatomy was a small part of the lecture. Due to the reduction in teaching time for Limb anatomy, this lecture could not be accommodated in the new program.

While the C21 curriculum was introduced in 2010, the overall curriculum had undergone significant changes for several years before that. A case study exercise on Sonic Hedgehog signaling had been part of the Cell Biology course (but led mostly by faculty involved in Embryology teaching) and was conceived as an integrated exercise combining embryonic development of the face, biochemistry of cholesterol synthesis, signaling pathways and human disease. A series of cases was discussed encompassing holoprosencephaly, Smith-Lemli-Opitz syndrome, and basal cell carcinoma – which all have a relationship to defective Shh signaling. The Cell Biology course was originally given in the second semester just prior to Histology and after Embryology and Biochemistry were completed. When Cell Biology was moved to the beginning of the academic year closer to molecular biology and biochemistry, the case study exercise was jettisoned; students did not have the appropriate background in embryology and course teaching hours were reduced.

METHODOLOGIES FOR TEACHING EMBRYOLOGY

Embryology is taught in many modalities at different institutions. Problem-Based Learning (PBL), Team-Based Learning (TBL), self-directed online modules, and computer animations are some of the more recent innovations. Indeed, given sufficient time in the curriculum, Embryology courses can be taught entirely in a case-based format where students prepare on their own using online modules and other materials provided by the instructors and participate in a small group discussion on each topic [4]. However, a recent survey of medical schools by the American Association of Medical Colleges confirmed that the vast majority of medical schools use lecture formats (www.aamc.org/curriculumreports). The number of hours devoted to Embryology has decreased and now averages about 18 [5], while ours is approximately 20. The issue we faced during the curriculum reform was how to define the philosophy of learning of Embryology and the essential core material given the time constraints, especially in light of the enormous amount of material being presented in the integrated modules on other subjects. Our total scheduled time for students for all subjects is limited to ~20 hours a week and ordinarily includes three case studies, laboratories, TBL exercises and/or small group exercises that require student preparation or extensive student time. Unlike PBL, where embryology can be integrated into some of the clinical problems and comprehensive lectures are frequently not part of the didactic program, the different topics in a curricular block are somewhat independent. From a logistical standpoint, this means that each topic within a module is vying for the same block of time for small group teaching. Other teaching modalities, such as TBL, may not require *small group* time per se, but do require significant student preparation. Therefore, Embryology, as is true of much of the C21 curriculum, is taught primarily through lectures. We have a total of four small group exercises (90 minutes each).

The first two small group exercises (<15 students/group) cover the Early Stages of Embryology (beginning with the blastocyst). They are led by faculty in the Developmental Genetics Program at NYUSOM. The first exercise uses visual materials, including computer animations (such as *Interactive Embryology* [6]) to illustrate gastrulation, neurulation, somite formation, and embryonic folding. The second session begins with a discussion centering around cases of caudal dysplasia, sacrococcygeal teratoma, and spina bifida/anencephaly.

This session then continues with an introduction to the principles of developmental biology using as examples the induction of the neural plate and neural crest formation. It is not productive in our view to have students learn the many growth and transcription factors that influence all of the steps in early development. Rather, we discuss one particular example in detail emphasizing the underlying molecular basis while defining induction and introducing the concepts of a morphogen, homeobox genes, and transcriptional programs that determine cell fate. With the explosion of information concerning regulation of development of various organs and tissues, it also is not possible to have students memorize the morphogens and transcription factors involved in each organ system. Instead, lecturers choose a few examples of inductive processes to emphasize in their presentations. These examples use the same governing principles, and even some of the same growth factors introduced in the early stages small group exercises and often relate to known genetic mutations in humans that disrupt normal development. The other two small group exercises are case studies. Our philosophy in designing case studies is to use them as much as possible to review congenital malformations of a particular group of structures as well as their underlying embryological defects and pathophysiology. We endeavor to make them broad and ambiguous, so students have to consider many options. Students are expected to participate and actively discuss options during the session. Faculty and fellows serve as facilitators, similar to PBL facilitators.

One of our case studies takes place in the Cardiovascular block on congenital heart disease. Students have two lectures and a laboratory prior to the case study. The first lecture covers normal heart development and begins with a video (Development of the Normal Heart [7]). Simple lesions of the heart such as atrial and ventricular septal defects are presented. The second lecture is given by a pediatric cardiologist. After an introduction to blood flow (pressure differentials, oxygen tension, etc.) in the fetal, neonatal and adult hearts under normal and pathological conditions, the lecturer presents students with a series of cases in an interactive "chalk talk" format on the whiteboard. Subsequently, students visit a hospital pathology laboratory and examine preserved hearts. Here they discuss the various congenital malformations they encounter with pathology residents and pediatric cardiology faculty and fellows. Students are given several days to prepare for the case study, which is a case of Tetralogy of Fallot (TOF) presented in two parts as more or less a detective story and led by pediatric cardiology faculty and fellows. The initial symptoms indicate that an open communication in the heart is likely but the child is lost to follow-up. When seen again, the pulmonary stenosis and cyanosis characteristic of more severe TOF are evident. Students are told to come to class with their own differential diagnosis based on the first set of symptoms. This discussion is followed by a further consideration of the cyanosis and arrival at a definitive diagnosis. Included in the clinical information provided are descriptions of heart murmurs as well as actual EKGs and echocardiograms. The students are in the process of learning how to interpret heart murmurs and how to read EKGs, so the case integrates across cardiovascular disciplines.

The second case study is in the Reproduction block and concerns a case of ambiguous genitalia. Prior to the case study, students are given with a lecture on genital development that includes a number of case descriptions (mostly of teenagers who fail to menstruate at puberty). The case study materials provide clinical information, but students must consider many different possibilities. In preparation for the case study itself, they are expected to review urogenital development and to generate a broad differential diagnosis. While the case turns out to be one of congenital adrenal hyperplasia, many disorders of sex development

(DSD) and the underlying embryological defects, are considered and discussed including androgen deficiency, androgen insensitivity syndrome, 5α-reductase deficiency and ovotestesticular DSD (a.k.a., true hermaphroditism). The students have just completed the Endocrine section of the module, so the case study is designed as a bridging exercise that integrates steroid metabolism with urogenital embryology (Reproduction). The small group sections are led by a pediatric urologist and pediatric endocrinologists, most of whom have treated intersex patients. Also discussed are options for surgical correction and issues of gender identity and assignment. A similar type of case study exercise using the example of a teenage patient who fails to menstruate has been developed and used as part of a case-based anatomy program at the Cleveland Clinic Lerner College of Medicine [4].

Student Resources

The availability of online and web-based resources is transforming Medical School teaching. Visual materials are particularly useful for learning Embryology. Some of the best online animations are the University of Indiana School of Medicine website (http://www.indiana.edu/~anat550/embryo_main/index.html) and, for congenital malformations of the heart, the Children's Hospital of Cincinnati (http://www.cincinnatichildrens.org/patients/child/health/).

In addition, many schools have developed their own materials. Some of these are available through MedEd portal on American Association of Medical Colleges website (https://www.mededportal.org/). We incorporate these and other computer animation programs and online videos into our teaching program. We also have produced a series of case-based online exercises for learning aspects of cardiovascular development in house (unpublished). In addition, all lectures are recorded so students are able to view and review them online using iTunes. A number of textbooks are commonly used in Embryology courses, including those by Moore et al. [8], Sadler [9], Carlson [10] and Schoenwolf et al. [11]. Shorter, streamlined versions of the Moore et al. [12] and Sadler [13] texts are also available. All of these texts are excellent, each with a different emphasis. The full version of Sadler tends to be shorter and less comprehensive than the others, while Schoenwolf et al. is the most detailed as far as developmental mechanisms are concerned. Moore et al. is very useful for a clinically oriented course and for many years we required it in our course.

However, none of these books are a perfect fit for our current Embryology material. For some topics, we have more advanced clinical information. In addition, we do not expect students to learn the complexities of the molecular aspects of the developmental pathways for all organ systems, while the textbook authors endeavor to be reasonably comprehensive in this regard. Therefore, we have written our own online "textbook" that is tailored for our course with input from our instructors (unpublished). Additionally, we require the Sadler textbook because it contains the figures referred to in the online chapters.

Assessment

A variety of methods may be used to assess Embryology learning by students. These range from traditional multiple choice questions to embryology questions in Gross Anatomy practical exams to TBL Readiness Assurance Process quizzes to quizzes given as part of online modules. Our assessment for Embryology is based on a combination of modalities, but we do not require students to identify embryonic structures. Although most of the assessment is in the form of multiple choice questions, we make an effort to integrate the Embryology with the rest of the material being presented. Moreover, while recall questions are part of our examination, application-type questions also are used. An example is given in Figure 1.

> Vascular endothelial growth factor receptor 1 (VEGFR-1) has a naturally occurring splice variant (sVEGFR-1) where a stop codon occurs near the DNA sequence encoding amino acid 760, just at the beginning of its transmembrane domain (aa 759-81). The syncitiotrophoblast produces both a full length and a truncated version of the protein. In pre-eclampsia, a condition that affects 10% of pregnant women, there is stunted growth and remodeling of maternal blood vessels during placenta formation. This eventually reduces the ability of the placenta to support normal fetal growth and can cause serious complications in the mother as well. Pre-eclampsia is correlated with an increase in the synthesis of sVEGFR-1 and its detection in the maternal bloodstream, leading some scientists to theorize that its overproduction causes an imbalance that is responsible for the placental pathology. Why does this theory make sense?
>
> A. sVEGFR-1 is required for the initial penetration of syncitiotrophoblast into maternal blood vessels.
> B. sVEGFR-1 inhibits fetal growth by preventing the development of an efficient fetal vascular system.
> C. sVEGFR-1 dimerizes with the normal transmembrane VEGFR-1 and activates it in the absence of VEGF.
> D. sVEGFR-1 dimerizes with the normal transmembrane VEGFR-1 and inhibits its response to VEGF.
> E. sVEGFR-1 binds to and sequesters VEGF, thereby antagonizing VEGF responses.
>
> *Answer: E*

Figure 1. Sample Application-type Multiple Choice Question Integrating Cell Biology and Embryology.

This question is one of a five question set on VEGF, its receptors and their role in signal transduction. Students are told that VEGF receptors are members of the EGF receptor signaling family, which they have studied. They also should know that antibodies to VEGF and its receptors are used to block angiogenesis in cancer patients.

These topics are part of the Core Foundations of Medicine (CFM) module in the Cell Biology segment (as noted above). In an Embryology lecture, students are introduced to the formation of the placenta and the role of the syncitiotrophoblast in invasion of maternal blood vessels, but they have not been taught about pre-eclampsia. They are asked to apply their knowledge to answer the questions. In addition to multiple choice questions, we use essay questions for most of the modules. For example, students may be asked to interpret a set of data concerning growth factor regulation of somite differentiation in the CFM module. After studying growth factors and learning about embryonic induction of the neural plate and neural crest, as noted above, they must use their knowledge of one inductive process to explain another. In the Cardiology module, the essay involves one of several cases – a cyanotic child, a child with congestive heart failure, etc. which are alternated year to year.

Students are asked to give a differential diagnosis based on the congenital abnormalities they studied. In the Endocrine/Reproduction module, we may ask students to consider why a “female” athlete with somewhat elevated testosterone might have a congenital malformation or error of metabolism that may result in her being banned from competition. Alternatively, we may ask why the drug finasteride (a.k.a., Proscar, Propecia) is used to treat benign prostatic hypertrophy and male pattern baldness and also why pregnant women are warned not to use the drug and even to handle the bottle. To answer these questions, which are given as part of more extensive clinical scenario, students must be able to apply their knowledge of embryology rather than just recall details, since no similar examples are discussed during the module itself. We administer a quiz at the beginning of small group exercises and case study sessions to ensure that students know the basic material.

Student preparation is critical to allow the case discussions to proceed at sophisticated level. These quiz grades are averaged in to the relevant module exam grades. By contrast, quizzes associated with online computer modules are not included in the grades because we encourage students to use any resources available to them to answer the questions while they are learning the material.

Problems Associated with Assessment

Assessment of Embryology in integrated formats can be problematic. In integrated Embryology and Gross Anatomy courses, Gross Anatomy questions far outnumber Embryology on the exams. As might be expected, students often tend to concentrate on Anatomy and spend considerably less time studying Embryology, especially in a pass-fail school such as ours. To deal with this issue, in the past we required students to pass Embryology as a separate component in order to pass the combined course. Now that Embryology is integrated throughout the curriculum, the limited contribution of Embryology to module material, and therefore its weight on examinations, becomes a very significant issue. In our renal segment, for example, there is a single lecture on renal and urinary development and malformations. While clinically oriented, this material accounts for <5% of the examination. On the other hand, a goal of our curriculum is to provide the student with an adequate background to become an independent, lifelong learner (Table 1). To some extent, it is the students’ responsibility to take advantage of the opportunities given to them to pursue a deeper comprehension and integration of the material.

Integration of Embryology with the Clinical Curricula

A key goal of the adopting the core competencies into the educational program was to promote integration of the basic sciences in the clinical training program throughout the entire four year curriculum. This has been a challenge at NYUSOM during the basic science years;

our Practice of Medicine (POM) course, where students learn clinical skills concurrently with their basic science courses, has also been reduced to 18 months. In POM, students attend lectures and discussions and shadow physicians. They follow a particular patient or patients throughout the course of their treatment, when possible. In addition, they visit hospitals, clinics, and test sites relevant to the areas they are studying in their organ system modules. One area of Embryology coordinated with POM has to do with the case study on Genital system development (see above). Complex ethical and psychological issues can arise for children with Disorders of Sex Development and their parents regarding gender assignment and gender identity. These issues are touched upon briefly in the Embryology case discussion, and also are taken up at approximately the same time in a unit of the POM course. Other related topics, such as sexual orientation, are explored and discussed with clinical faculty in this segment of POM as well. Pre-clerkship Embryology is also revisited during the Obstetrics/Gynecology (Ob/Gyn) and Pediatrics clerkships, which are required of all students. In Ob/Gyn, there are units on prenatal testing, fetal ultrasound, fetal lung maturation, high risk pregnancy, ectopic pregnancy, normal and abnormal placentation, reproductive choice, infertility, and assisted reproductive methods. All of these topics are also covered in the basic science organ system modules (primarily Reproduction). Lectures, hospital and clinic visits (including our renowned *in vitro* fertilization clinic) and shadowing of physicians are part of this clerkship, as is assisting in actual deliveries. In Pediatrics, students visit neonatal intensive care units and gain experience in pediatric cardiology, both through lectures and through attendance at the pediatric cardiology clinic. As a major medical center, NYUSOM, together with Bellevue Hospital, our sister institution, receives many referrals and students invariably will encounter newborns with malformations they learn about in Embryology.

Interested students also have the option to sign up for a four-week Pediatric Surgery elective during their third or fourth years. This elective is one of several offered concurrently by the Department of Surgery along with oncological, vascular, and endocrine surgery. Students shadow Pediatric Surgeons and observe operations, frequently on infants with congenital malformations. A component of the Pediatric Surgery elective is a tutorial with two faculty members (a neonatologist interested in pulmonary development and myself) familiar with congenital malformations. Students select basic science or translational medicine articles from the literature on a specific malformation and explore such topics as molecular mechanisms of development in mouse models or the genetic basis of the defect in human fetuses. Students meet individually with the faculty members to discuss the articles. Subsequently they present one of the articles at a Journal Club attended by the faculty and their student colleagues of all the Surgery electives.

It is apparent to the faculty that in this self-selected cohort at least, NYUSOM students entering the elective (as opposed to others come from other local institutions that lack formal embryology in their basic science programs) are well-grounded in the basic molecular biology and embryology they need to critically evaluate current literature in the field.

Conclusion

Embryology may be successfully integrated into an organ system-based curriculum. A systems-based approach lends itself to in-depth teaching and learning of congenital malformations and their clinical correlations. By incorporating the gastrointestinal and head and neck portions of Embryology with Gross Anatomy, we preserved some of the advantages of teaching the two disciplines together as an integrated block. The reduction in teaching hours impacted Embryology as well as the other courses. Nonetheless, we believe that students are receiving a solid foundation in the field that enables them to function at a high level in their obstetrics and pediatrics clinical rotations and will serve them more than adequately in their future medical careers.

REFERENCES

[1] Carlson BM. Embryology in the medical curriculum. *Anat, Rec.* 2002 Apr 15;269(2):89-98.

[2] Educational Affairs Committee A. A clinical anatomy curriculum for the medical student of the 21st century: developmental anatomy. *Clin. Anat.* 2000;13(1):17-35.

[3] Triola MM, Holloway WJ. Enhanced virtual microscopy for collaborative education. *BMC Med. Educ.* 2011 Jan 26;11:4.

[4] Drake RL. A unique, innovative, and clinically oriented approach to anatomy education. *Acad. Med.* 2007 May;82(5):475-8.

[5] Drake RL, McBride JM, Lachman N, Pawlina W. Medical education in the anatomical sciences: the winds of change continue to blow. *Anat. Sci. Educ.* 2009 Nov-Dec;2(6):253-9.

[6] Lash J. Interactive Embryology: The Human Embryo Program CD. Sunderland, MA: *Sinauer Associates*; 1998.

[7] Rushmer RF, Blandau RJ. Congenital malformations of the heart; Pt. 1: *Development of the normal heart.* Seattle, WA: University of Washington School of Medicine; 1953.

[8] Moore KL, Persaud TVN, *Torchia MG. The developing human : clinically oriented embryology*. 9th ed. Philadelphia, PA: Saunders/Elsevier; 2013.

[9] Sadler TW, Langman J. Langman's medical embryology. 11th ed. Philadelphia: Wolters Kluwer Lippincott Williams and Wilkins; 2010.

[10] Carlson BM. *Human embryology and developmental biology*. 4th ed. Philadelphia, PA: Mosby/Elsevier; 2009.

[11] Schoenwolf GC, Bleyl SB, Brauer PR, Francis-West PH. *Larsen's human embryology.* 4th ed. Philadelphia, PA: Elsevier/Churchill Livingstone; 2009.

[12] Moore KL, Persaud TVN, Torchia MG. Before we are born : *essentials of embryology and birth defects*. 7th ed. Philadelphia, PA: Saunders/Elsevier; 2008.

[13] Sadler TW, Langman J. *Langman's essential medical embryology.* Philadelphia: Lippincott Williams and Wilkins; 2004.

In: Education in Anatomical Sciences
Editor: Paul Ganguly
ISBN: 978-1-62618-488-6

Chapter 10

EDUCATION IN ANATOMICAL SCIENCES - NEUROANATOMY: AN EVOLVING CURRICULUM CATERS TOWARDS PATIENT CARE

Ahmed Yaqinuddin
College of Medicine, Alfaisal University, Riyadh, KSA, Saudi Arabia

ABSTRACT

Teaching of anatomy is widely recognised as a corner stone in the education of physicians. In recent years, with an expanding curriculum, there has been increasing pressure on teachers of anatomy, specifically neuroanatomy, to reduce the number of teaching hours. Moreover, with the evolution of thought in the field of medical education and subsequent development of new models of medical curricula, there has been a shift in how the pedagogy of neuroanatomy is perceived? This chapter describes the neuroanatomy curricula in a traditional and a problem based curriculum setting. It discusses how a neuroanatomy curriculum can be modified so that it will increase retention of knowledge in the senior years of medical school. Finally, as there is a widespread adoption of outcome and competency based curricula, the future neuroanatomy curricula, not only need to be directed towards improving patient care, but also need to focus on developing specific competencies in future physicians.

1. INTRODUCTION

The study of anatomy has a history unlike many other modern-day disciplines in medicine; dating back to the times of Aristotle and Galen. This is the reason why changes in society and modern medicine have an impact on how anatomy is now being taught [1]. In United States, medical education is a cultural transfer from Europe [2]. This cultural transfer was made in three phases. In the first phase, physicians trained in Europe migrated to United States. In second phase, they inspired young prospective medical students in the United States (U.S.) to travel to Europe to acquire training. Finally, the newly trained physicians from

Europe returned to the U.S. and established medical schools there. In the early 1900's, in the United States, reforms were made in the medical curriculum which led to conversion from an apprenticeship model to a discipline- based approach [3]. These reforms were enforced by the faculty which was trained in Europe [4]. Franklin Paine Mall, an anatomist, was one of the most influential reformists in anatomical education [2]. During his studies in Germany, under supervision of embryologist Wilhelm His and physiologist Carl Ludwig, he learned the principles of medical education. He started to understand the value of learning anatomy by "doing". This is the main learning strategy which has been in practice in anatomy teaching since the first human dissections in Alexandria by Erasistratus and Herphilus [5], animal dissections by Galen, and the emergence of anatomical enquiry during the renaissance in Europe [6]. Mall was fully convinced that the role of Anatomy professors in medical education is to perform dissections with students rather than using lectures as the main method of teaching [7]. He described his experience in following words "I have learned to teach anatomy with few lectures, it is not difficult and the success is great" [2]. Mall also learned from his German mentors that learning anatomy is not only about clinical application, but rather application of anatomical knowledge in biomedical research. [8] These potential avenues of research in anatomy as identified by Mall are currently referred to as "Cell biology" and "Developmental Sciences". In 1893, Mall became the chairman of the Anatomy Department at John Hopkins Medical School. Here, he introduced a program of anatomy which had two main thrusts 1) scientific scholarship in anatomy and 2) teaching anatomy through dissections. In-order to achieve these goals he utilized highly productive researchers to teach anatomy through dissections [7-9]. In his new program, he proposed only essential anatomy to be taught to all medical students and elective courses should be available to students who have a desire to acquire more deep anatomical knowledge [7]. Mall expected that these new arrangements would allow the teachers to have more time devoted to their research, which he thought was absolutely essential for vitality of this science [7].

Anatomy Departments in United States maintained their leadership role as the focus of research moved from tissue to cell, and eventually to the molecular level by the second half of the 20th century [10]. With the sudden increase in knowledge and the integration of cellular, molecular and physiological mechanisms to understand the function of human body, the discipline-based approach soon became unsuitable for education of future physicians [11]. In the 1950's, the organ-system based approach was adopted, where the curriculum content was developed by newly formed module (topic) committees, rather than disciplines or departments. The concept was further extended in newer curricula, such as, problem-based and clinical-presentation based learning which attempted to integrate anatomy with clinical and applied aspects of disease [3]. This approach was adopted based on cognitive studies assessing how knowledge is acquired.

The changes in the medical curriculum brought about significant changes in the teaching method and time allotted to all anatomical disciplines including gross anatomy, histology, embryology and neuroanatomy. In the 1960's, the total time dedicated to gross anatomy teaching was 248 hours which was reduced to 189 hours in 1973 and now it is 149 hours on national average in United States [12, 13]. Similarly, the time dedicated to neuroanatomy has also significantly dropped with total time allocated to it was 90 hours in 1960 which dropped to 70 hours in 2001 [13] additional 18% drop was seen during 2002-2009 in US medical schools [12]. In United Kingdom, a survey of 21 medical schools was conducted in 1999-2000 to explore the total number of hours dedicated to gross anatomy, neuroanatomy, embryology and histology teaching. Among the schools surveyed 12 followed organ-system based, 4 followed problem based learning and 5 traditional (discipline) based curricula. Surprisingly, this survey showed that neuroanatomy was allocated more time in a PBL system (30 hours) as compared to an organ system based curriculum (20 hours) [14].

Gross anatomy is one of the main bodies of knowledge which must be acquired in order to apply diagnostic and treatment skills required for clinical competence [12]. Similarly, basic knowledge of neuroanatomy must be attained to have the skills required to practice clinical neuroscience [15]. This is highlighted by one text book which states that "Neuroanatomy is the corner stone upon which is built an understanding of nervous system and its disorders" [16].

2. Neuroanatomy Curricula

Neuroanatomy is taught in every medical school around the globe. In some places it is taught as a standalone course; at others, it is taught integrated with neurophysiology, neurophramacology and neuropathology in organ-system based blocks. Alternatively, neuroanatomy is combined with "Behavioral sciences" or "Gross Anatomy of Head and Neck" [15, 17]. In the following text, we will describe two examples of neuroanatomy curricula in different settings.

I. Neuroanatomy Curriculum at College of Medicine, University of Cincinnati

The neuroanatomy curriculum at the College of Medicine, University of Cincinnati, before 1998 used to run in the first year in a traditional format. In this curriculum, neuroscience course (Neuroscience-1) was preceded by a 7 week long, lecture and laboratory based course covering the Gross Anatomy of Head and Neck region [17]. The titles of the lectures and laboratory sessions, prior to reorganizing into organ-system block based curriculum, are presented in Table 1.

A new curriculum was adopted in the fall of 1998. The traditional gross anatomy, physiology, biochemistry, microscopic anatomy and neuroanatomy courses were integrated to form organ-system based modules [17]. Neuroanatomy, which was previously taught as a part of two separate courses " Gross anatomy of Head and Neck" and "Neuroscience -1", was merged to form a new course called "Brain and Behavior-1"[17]. The Brain and Behavior-I was a 10 week long integrated multidisciplinary neuroscience course which was part of the 4-year long longitudinal neuroscience curriculum of the medical school [17]. This course was comprised of integrated lectures and laboratory sessions in neuroanatomy, neurophysiology, neuropharmacology, behavior sciences and psychiatry. The second course in this series "Brain and Behavior II" was offered to the students in Year-2 [17]. This course was comprised of integrated core disciplines of neuropathology, neuropharmacology, behavior sciences and psychiatry.

Some examples to illustrate how the integration between Gross Anatomy of Head and Neck course and Neuroscience-I course were achieved in the new offering of Brain and Behavior-I was depicted in Table-2 [17].

Table 1. List of topics prior to the integration of Head and Neck block with Neuroscience-1

Head and Neck Anatomy	**Neuroscience-1**
Cranial vault and anatomy of base of brain	Neurobiology
Cranial nerves	Neurotransmitters and neuronal communication
Head and superficial face	Gross anatomy of brain and spinal cord
Eye and orbit	The vascular system of brain
Ear	The sensory system
Parotid gland and region	The motor system
Temporal and infratemporal fossae	The visual and auditory system
Triangles of neck	Chemical senses
Anatomy of Pharynx and larynx	The reticular activating system and consciousness
Nose and Paranasal sinuses	The oculomotor system
Pterygopalatine fossa	Memory system
Oral cavity and palate	Integrative systems

* Adapted from Giffin and Drake, Anatomical Record, 2000.

Table 2. Examples of integration of Gross Anatomy of Head and Neck with Neuroscience-1 in the new course: "Brain and Behaviour-1"

Brain and Behavior-1	
Gross Anatomy of Head and Neck	Neuroscience-1
Example:1 Lecture: Meninges and cranial cavity Lab: Meninges and interior of skull	Example:1 Lecture: Orientation and organization of CNS I and II
Example-2 Lecture: The cranial fossae	Example-2 Lecture: The cerebrospinal fluid system Lecture: Neuroimaging
Example-3 Lecture: Orbit Lab: Orbit and its content	Example-3 Lecture: Retina Lecture: Central retinal pathway Lecture: Visual physiology Lecture: Neural control of eye movement and balance Clinical lecture: Eye ball movement Clinical lecture: Dizzy patient Case based discussion: Visual, oculomotor and balance.

* These selected examples demonstrate how Gross Anatomy of Head and Neck was integrated with Neuroscience-1 course. The table is adapted from Giffin and Drake, Anatomical Record, 2000.

The Brain Behavior-I course was set to have three examinations during its 10 week run that would test the knowledge of core disciplines in an integrated manner. The integration of basic science with clinical material was tested by developing questions using clinical vignettes [17]. Since students who took the new Brain and Behavior-I course had not been introduced to cranial nerve anatomy, emphasis in this course was given to the teaching of cranial nerve anatomy. An introductory lecture introduced the students to functional components of the cranial nerves. It also, introduced students to the relationship of cranial nerves with parasympathetic ganglia and autonomic nervous system. This lecture was followed by two more lectures on cranial nerve anatomy [17].

The second key change in this integrated Brain and Behavior-I course was the use of case-based sessions in a small group setting [17].There were five goals which were achieved through the introduction of case based sessions:

1. Strong clinical integration with the basic knowledge of neuroscience was developed through careful construction of cases such that analysis of these problems reinforced basic science knowledge.
2. Development of critical thinking among students and development of a practical approach to problem solving.
3. Development of skills in using appropriate resources (information technology).
4. Interpersonal skills were developed by working in teams to solve problems.
5. Written and verbal communication skills were developed, since each group was required to discuss cases in small group settings and then submit a written document describing their analyses of the cases.

The strategy of "focus questions" was used to achieve learning objectives in the case-based sessions [17]. The students were required to answer a series of focus questions related to the case in their written reports. These questions were categorized in four sections including anatomy/clinical presentation, pathophysiology and cellular/molecular aspects, clinical laboratory data, and prognosis and treatment options [17].

As described by Griffin and Drake, this course was well received by both students and tutors. Faculty and student feedback showed that this course led to considerable improvement in students verbal and written skills [17].

II. Neuroanatomy in a Problem Based Learning Curriculum, at College of Medicine, Alfaisal University

The College of Medicine (CoM) at Alfaisal University was founded in 2008 and has adopted a PBL-based hybrid curriculum. This curriculum is taught over three successive phases over the period of five years. The college has an integrated longitudinal neuroscience curriculum with neuroscience being taught in all three phases of this curriculum. In Phase-1 of the curriculum, the first neuroscience course is introduced as a structure and function integrated block of eight weeks. There is a second neuroscience block of 6 weeks in the Phase-2 of the curriculum which focuses on neuropathology, neuropharmacology and psychiatry. Finally, the students are exposed to neuroscience block of 8 weeks in the clerkship phase, where they learn basic principles of the practice of neurology, neurosurgery, paediatric neurology and psychiatry.

The first neuroscience block in Phase-1 of the curriculum is 8 week long with each week being assigned a specific theme. Every week there is a Problem based learning (PBL) case which is constructed to match the theme of the week.

Table 3. The list of themes and matched PBL cases

Weeks	Themes	Problem based learning cases
1	Neuron: The functional unit of the nervous system	Walking with Difficulty (Guillain barre syndrome)
2	Motor and sensory system: spinal cord and Tracts	A pain in the neck (Syringomyelia)
3	Motor and sensory system: brain stem	Vertigo (Lateral medullary syndrome)
4	Motor and sensory system: cerebrum	Hemiplegia (Stroke)
5	Motor modulation : basal ganglia and cerebellum	Tremors (Parkinsonism)
6	Sensory modulation and homeostatic functions: thalamus, hypothalamus and reticular formation.	Coma (Stroke)
7	Complex brain functions: sleep, emotions and memory	Seizures (Epilepsy)
8	Meninges and ventricular system	Stiff neck (Meningitis)

a) Lectures

This course has 15 anatomy lectures and 18 physiology lectures. There are 6 structured 2 hours long anatomy laboratory sessions and 2 electrophysiology labs. In-addition, there are 6 large group discussion sessions where discussion revolves around brief clinical cases and focused questions approach is utilized. The complete list of titles and objectives of lectures is given in Table 4.

Table 4. Titles and objectives of lectures

Week 1	Neuroanatomy lectures	Neurophysiology lectures
	Neuron, synapse and glial cells **Objectives:** Decribe the structure and function of neuron Describe the structure and function of synapse Describe the principles underlying	**Overview of the nervous system** **Objectives:** Learn the general organization of the nervous system into the Central (consisting of the Spinal Cord and Brain), and the Peripheral (consisting of the Somatic and Autonomic) Nervous System. **Electrical Properties of Excitable Cells – Resting membrane potential**

Week 1	Neuroanatomy lectures	Neurophysiology lectures
	successful and unsccessful regeneration in the PNS and CNS Describe the structure and function of astrocytes, oligodendrocytes, microglia, ependymal cells and Schwan cells	**Objectives:** Explain why a cell has an electrical charge difference (voltage) across its membrane. Using the Nernst equation, explain the effects of altering either the intracellular or extracellular Na+, K+, Cl-, or Ca2+ concentration on the equilibrium potential for that ion. Explain the effect of changing extracellular concentrations of Na+, K+, Cl-, or Ca2+ on the resting membrane potential (and excitability of a neuron) **Synaptic Neurotransmission & Neural Communication** **Objectives:** Describe the main morphologic features of synapses. Compare electrical and chemical synapses transmission based on velocity of conduction, fidelity, and the possibility for neuromodulation (facilitation or inhibition). Contrast the terms temporal summation and spatial summation at the chemical synapse, Explain the sequence of events during chemical synaptic transmission Describe fast and slow excitatory and inhibitory postsynaptic potentials, outline the ionic fluxes that underlie them, and explain how the potentials interact to generate action potentials. Describe the ionic basis for inhibitory and excitatory post-synaptic potentials and how these changes can alter synaptic transmission. Define convergence and divergence in neural networks, and discuss their implications. understand the concept of synaptic integration and plasticity

Table 4. (Continued)

Weeks 2	Neuroanatomy lectures	Neurophysiology lectures
	Spinal Cord & Tracts 1, 2 and 3: **Objectives:** Describe the relationship of segmentation of the spinal cord and spinal nerves to segmentation of the vertebral column. Describe the significance of dermatomes and myotomes. Describe the general cellular organization of the spinal gray regions concerned with somatosensory, motor and autonomic functions Describe the functional organization of the white matter into lateral, dorsal, and ventral Describe the functional significance of the dorsal root ganglia, dorsal roots, ventral roots, and mixed spinal nerves. Correlate afferent fiber types with somatosensory modalities. Describe the structural and functional differences between the dorsal column/ medial lemniscal and the anterolateral systems. Describe the constituents and pathways of descending motor systems in the spinal cord. Describe the structural and functional correlates of the differences between upper and lower motor neuronal populations.	**Reflexes, Muscle Tone:** **Objectives:** Understand the general organization of the motor system. Explain the structure and function of the muscle spindle and Golgi tendon organ. Trace the neuronal activity initiated by striking the patellar tendon with a percussion hammer (the patellar tendon reflex) that leads to contraction of a muscle. Contrast this reflex with the inverse myotactic reflex. Describe the role of the gamma efferent system in the stretch reflex, and explain the significance of alpha gamma co-activation. Describe the properties of the flexor reflex initiated by touching a hot stove. Identify when pain is sensed, when flexor contraction occurs, and the neuronal connections and role of the crossed extensor reflex. Understand the basis of common tendon/stretch reflexes and their utility in clinical diagnosis Identify the spinal levels of common tendon reflexes **Sensory Receptors** **Objectives:** Understand the use of the term receptor in sensory physiology. Understand the basic principles of sensory transduction Understand how sensory receptors code stimulus intensity and duration. Understand how specific sensory receptors are associated with specific CNS sensory pathways. Classify sensory receptors according to their stimulus modality. Define and understand the concepts of adequate stimulus,doctrine of specific nerve energies, and receptive field Describe the following cutaneous and proprioceptive mechanoreceptors and their function: Pacinian corpuscles, Meissner's corpuscles, Ruffini endings, Merkel cell, A-delta and C free nerve endings, Golgi tendon organ, muscle spindle

Weeks 2	Neuroanatomy lectures	Neurophysiology lectures
		Mechanism of Pain: **Objectives** Define pain Identify the receptor endings and fibre types involved in pain perception Understand the range of stimuli inducing pain. List commonly understood chemical mediators of pain and their mechanism of transduction. Identify the ascending pain pathways and their termination in the brain. Explain the organization and function of the body's endogenous analgesia system
Week 3	**Neuroanatomy lectures**	**Neurophysiology lectures**
	Brain stem 1, 2 and 3 **Objectives:** Describe the principal structures of the brainstem: major pathways and cranial nerve nuclei. Correlate anatomical features with the functional properties of the major pathways and cranial nerve nuclei. Describe the functional consequences of brainstem injury. Describe the brainstem control of eye movements; relationships with vestibular functions. Describe the role of the reticular formation in respiratory and cardiovascular control and consciousness. Describe the brainstem control of eye movements; relationships with vestibular functions.	**Brain Stem Reflexes** **Objectives:** Describe the function of the following brain stem reflexes: cardiovascular baroreceptor, respiratory stretch receptor. Contrast the effects of intra-axial and extra-axial brain stem lesions **Neurotransmitter Systems** **Objectives:** Define the characteristics of a neurotransmitter. Learn the synthetic pathways, inactivation mechanisms and neurochemical anatomy and mechanisms of receptor transduction for the following neurotransmitters: 1. Catecholamines (DA, NE, E) 2. Acetylcholine (ACh) 3. Serotonin (5-hydroxytryptamine; 5-HT) 4. Histamine 5. GABA (gamma-aminobutyric acid) 6. Glutamate 7. Endorphins 8. Enkephalins 9. Dynorphins 10. Substance P Learn the major receptor classifications and representative receptor agonists and antagonists for the above transmitters. Describe the relationships between neurotransmitter dysfunction and neuropathology.

Table 4. (Continued)

Week 4	**Neuroanatomy lectures**	**Neurophysiology lectures**
	Blood supply of brain 1 and 2 **Objectives:** Describe the major neural structures served by the anterior, middle and posterior cerebral arteries; Describe the major vessels that supply the brain stem, cerebellum and spinal cord. Describe the origin, course and distribution of the internal carotid artery and its major branches Describe the major features of the arterial circle(Willis) Describe the course of the vertebral arteries and their entrance through the foramen magnum. Describe the distribution pattern of the major branches of the basilar and posterior cerebral arteries Describe the major features of the venous drainage of the brain. Discuss the relationships of the major cerebral veins. Describe the major venous sinuses and their tributaries	**Regulation of Cerebral Blood Flow** **Objectives:** Understand how cerebral blood flow is measured Describe the factors affecting brain blood flow Outline the effect of intracranial pressure changes on cerebral and systemic circulation Understand brain metabolism and oxygen requirements. **Functional Areas of the Brain** **Objectives:** Identify the primary sensory and motor areas and the effect of stimulation or lessoning Identify the secondary sensory and motor (unimodal association) areas and the effect of stimulation or lesioning. Describe the functions of the association cortex.
Week 5	**Neuroanatomy lectures**	**Neurophysiology lectures**
	Basal Ganglia **Objectives:** Describe the anatomical features of the basal ganglia Describe how the cerebral cortex, basal ganglia and motor thalamus are interconnected. Describe the connections and loops of basal ganglia **Cerebellum** **Objectives:**	**Basal Ganglia functions:** **Objectives:** Describe the major interconnections between components of the basal ganglia and the motor cortex. Identify the neurotransmitters determining the flow of information in the system. Describe the overall function of the basal ganglia in movement control and initiation in association with medial and lateral motor systems. List the appropriate signs of rigidity, dyskinesias, akinesia, and tremor for Parkinsonism, chorea, hemiballism, and athetosis. Assign a likely lesion site or chemical system defect for each clinical syndrome.

Week 5	Neuroanatomy lectures	Neurophysiology lectures
	Describe the histological features of cerebellum. Describe how cerebellum influences the function of the motor cortex via its projections through the thalamus. Recognize the effects of unilateral cerebellar lesions	Describe the rationale for treatment of Parkinsonism with anticholinergic drugs, l-DOPA, or transplantation of catecholamine-producing cells. **Cerebellum functions:** **Objectives:** Describe the roles of the cerebellum in the regulation of skilled movement. List three functional divisions of the cerebellum, detailing the input and output connections of each. Be able to differentiate the functions of each and their integration with lateral and medial motor systems. On the basis of input-output organization, somatotopic organization, and overall function, predict the neurological disturbances that can result from disease or damage in different regions of the cerebellum
Week 6	**Neuroanatomy lectures**	**Neurophysiology lectures**
	Thalamus **Objectives:** Describe the organization of the thalamus with respect to its reception and processing of afferent information from sensory and motor systems and the linkage of functionally defined regions of the thalamus with functional areas of the cerebral cortex. **Development of Nervous System** **Objectives:** Describe the longitudinal organization of the CNS in relation to the development and maturation of the neural tube	**Hypothalamus and Homeostatic regulation** **Objectives:** General homeostatic functions of the hypothalamus – the hypothalamus receives signals from many sources in the nervous system Role of the hypothalamus in temperature regulation Role of the hypothalamus in the regulation of plasma osmolality – neurohypophysis and antidiuretic hormone (ADH). Role of hypothalamus in reproduction – female climax, parturition and milk let down (oxytocin) Role of hypothalamus in circadian rhythm. Role of hypothalamus in circadian feeding. Integration of thalamus and hypothalamus **Reticular activating system and Consciousness** **Objectives:** Define consciousness Describe the neural systems important for the regulation of sleep-waking. Describe the neurochemical systems important for sleep and waking. Describe the mechanisms important in the production of coma.

Table 4. (Continued)

Week 7	**Neuroanatomy lectures**	**Neurophysiology lectures**
	Limbic System **Objectives:** Describe the limbic lobe and papez circuit. Describe connections of hippocampus and amygdala. Describe the function of hippocampus and amygdala	**Normal brain waves** **Objectives:** Describe the origin of spontaneous electrical activity of the cerebral cortex. Describe the primary types of rhythms that make up the EEG and the behavioral states that correlate with each. Distinguish EEG activity from evoked potentials and the uses of evoked potentials. **Abnormal cortical electrical activity: Seizures** **Objectives:** Recognize normal and abnormal EEG records. Describe characteristics of generalized and partial seizures. **Memory** **Objectives:** List the parts of the brain that appear to be involved in memory in mammals, and summarize the proposed role of each in memory processing and storage. Temporal and qualitative categorization of human memory Explain the role of NMDA receptor activation in encoding memory Explain the mechanisms proposed for short term and long term memory storage.
Week 8	**Neuroanatomy lectures**	**Neurophysiology lectures**
	Meninges & Ventricular System **Objectives:** Describe the anatomical basis of the blood-brain barrier and the blood-csf barrier Describe the structural relationships of the duramater, arachnoid, and pia mater Describe the subarachnoid space; understand theconcept of cisterns around the brain	**Cerebrospinal fluid production** **Objectives:** Describe formation and reabsorption of cerebral spinal fluid, including the anatomy and function of the choroid plexi. Describe the normal pressure, volume, and composition of the CSF. Describe how CSF can vary in certain pathological conditions. Describe the endothelial basis of the blood-brain barrier, and predict the consequence of this barrier for the central

Week 8	Neuroanatomy lectures	Neurophysiology lectures
	Describe the configurations and relationships of the lateral ventricles, interventricular foramina, third ventricle, cerebral aqueduct, fourth ventricle, and the foramina of Luschka and Magendie. Describe the relationship of the meninges to the spinal cord and the importance of the lumbarcistern.	nervous system distribution of intravenously administered hydrophilic and hydrophobic drugs

b) Neuroanatomy Laboratory Sessions

In the anatomy laboratory sessions, teaching is conducted through the use of: (1) cadaveric prosections, (2) plastic models, (3) radiological and histological images, and (4) clinical cases. As previously described [18], Alfaisal, CoM has developed a unique approach called integrated resource sessions (IRS) in which students are required to rotate through wet and dry stations during anatomy practical sessions. In the dry laboratory session students are divided into four groups of 4-5 students each. Each group is given two-three short cases related to the theme of the week to work through over a period of 30 minutes with focused questions to answer. The students have to use labelled plastic models, CT/MRI, brain cross sections and angiograms to answer three types of questions 1) Locate the site of lesion on the model or image provided 2) relate the neurological signs in a given case with damage/injury to parts of nervous system 3) describe/define the pathology of the neurological disorder presented in the case. Two examples of typical stations are provided below as Case-1and 2 (Figure 1 and 2). In the wet laboratory stations students are asked to dissect different part of the brain related to the theme of the week and they are demonstrated structures using wet prosected brain specimens.

Case 1

A 20 year old woman, who had suffered from endocarditis, suddenly fainted and remained unconscious for several hours. On awakening, she was unable to speak. She was able to print words to form sentences with her left hand, but not with her right hand, which was flaccid and paralyzed. Several months later, loss of speech persisted and she showed a spastic weakness of the right arm and hand with increased resistance to passive movement and exaggerated tendon reflexes. The lower facial muscles on the right side were paralyzed.

Questions

1. *Locate the lesion on the plastic model provided, name the structures involved and which abnormality is associated with which structure.*
   a. Level: (A) Cortex (Primary motor cortex)
   b. Loss of speech: Broca's area involved
   c. Spastic weakness of right hand and weakness of right lower facial muscles: upper motor neuron lesion involving corticospinal spinal tract originating from primary motor cotex (Ventral part of pre central gyrus)

2. *Identify the artery involved on the normal angiogram provided:*
   (B) middle cerebral artery
3. *Define aphasia*
   Aphasia is the inability to understand or communicate speech, writing or signs.

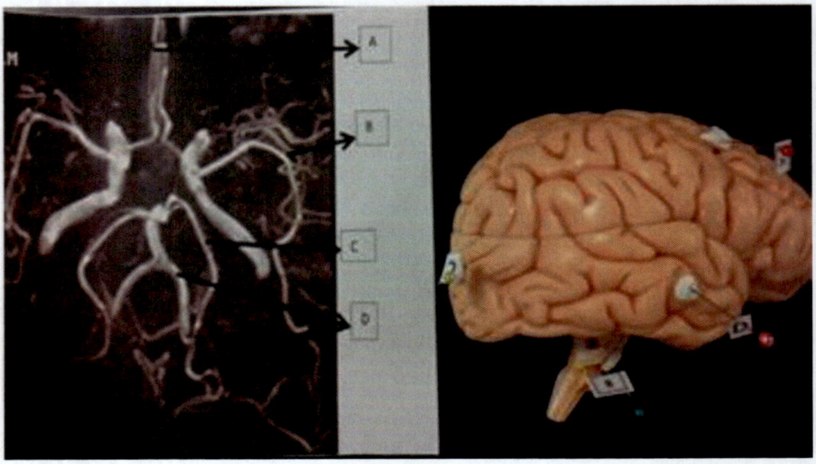

Case 2

A middle aged woman appeared in the clinic because of difficulties in walking and a sagging of the corner of the mouth. Her history showed that these abnormalities were the latest in the long series of events. About 5 years previously, the woman had a series of dizzy spells and complained of tinnitus in the right ear. Somewhat later she found it difficult to close her eyes tightly, and the corner of her mouth on the right side began to droop and did not rise when she smiled. Recently she experienced intermittent painful sensations in the right side of face, and now it become numb. Within the past several weeks, she noticed a tendency to sway to the right while walking. These recent events have been accompanied by difficulty in swallowing and hoarseness. Neurologic examination revealed a loss of taste on the right side of her tongue. No corneal reflex can be elicited on right side.

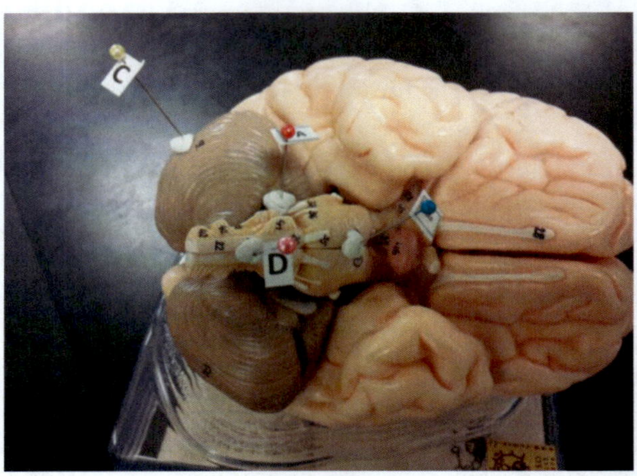

Questions

1. *Locate the lesion on the labelled specimen provided:* Cerebello-pontine angle
2. *Name the structures involved and specify which abnormality is associated with each.*
   a. *Dizziness:* vestibular nerve
   b. *Tinnitus:* cochlear nerve
   c. *Facial paralysis:* Facial nerve
   d. *Face pain and numbness:* Trigeminothalamic tract
   e. *Ataxia:* inferior cerebellar peduncle
   f. *Hoarsness:* Glossopharyngeal nerve
   g. *Taste:* facial nerve
   h. *Loss of corneal reflex:* facial nerve
3. *What is the probable etiology?*

A tumor pressing the nerves and structures around the cerebello-pontine angle.

c) Assessment

There are three methods of assessment which are used to test the objectives of this course. These include Multiple Choice Questions (MCQs), Short Answer Questions (SAQ) and Objective Structured Practical Examination (OSPE). The preparation of the examination starts with designing a blue-print of the examination which describes how the objectives of the course will be tested. The blue print of examination is usually developed by the block director to match the course objectives with the assessment modalities. According to the blue-print, questions are collected from the resource persons during every week of the course and are reviewed by a multidisciplinary question review committee which includes basic scientists and clinicians. An example of the blue-print of neuroscience examination is given in Table-5.

The neuroscience examination is comprised of 75 MCQs. These MCQs are constructed using clinical vignettes. The MCQs are carefully designed to test the objectives of lectures and large group discussions. The SAQs are developed such that they test objectives which are covered in the PBL sessions. Each SAQ has a brief clinical case and students are asked to answer a series of basic science questions relating clinical signs of the case with basic knowledge of Neuroanatomy and Neurophysiology. The OSPE component of examination (usually 20 stations, 1.5 minutes each) is developed in accordance with the blue-print. These OSPE stations test objectives which were covered during the Neuroanatomy and Electrophysiology laboratory sessions.

d. Conclusion

The Neuroscience course at Al Faisal, College of Medicine, has been running successfully with positive feedback from both students and faculty. However, in order to achieve a higher level of integration and clinical relevance, a proposal has been put forth to integrate Neuroscience course in Phase-2, which deals with Neuropathology, Neuropharmacology with this Basic Neuroscience course (Neuroanatomy and Neurophysiology) in the future.

Table 5. Blue-print of Neuroscience Examination

Themes/Disciplines	Anatomy	Embryology	Histology	Physiology	Total
Week-1 Neuron, The functional unit of nervous system	Neuron synapse and glial cells- 2 MCQs 1 OSPE station			Organization of nervous system-3 MCQs Electrical properties of excitable cells- 3 MCQs Nerve action potential- 2 MCQ Synaptic neurotransmission- 3-MCQ	13 MCQs and 1 OSPE stations
Week-2 Motor and sensory system: Spinal cord and Tracts	Spinal Cord structure and tracts- 4 MCQ 1-SAQ and 1-OSPE station		Spinal cord- 1 OSPE station	Reflexes and muscle tone- 2MCQ Somatic sensory receptors- 2 MCQ Pain Mechanism-2 MCQ EMG lab (1 OSPE station)	10 MCQs, 1 SAQ and 3 OSPE stations
Motor and sensory system: Brain stem	Brain stem structure and cranial nerve nuclei- 9 MCQs 1-SAQ and 5 OSPE stations			Neurotransmitter system 3 MCQ Brain stem reflexes- 2 MCQ	14 MCQ, 1 SAQ and 5 OSPE stations
Motor and sensory system: cerebrum	Cerebrum-3MCQ Blood Supply of Brain- 5 MCQ 2 OSPE Stations 1 SAQ		Histology of cerebrum 1 OSPE station	Regulation of cerebral blood flow- 2 MCQ Control of voluntary movements- 2 MCQs Speech mechanism- 2 MCQs	14 MCQ, 1SAQ and 3 OSPE stations
Motor Modulation : Basal Ganglia and Cerebellum	Basal Ganglia- 2 MCQ, 2OSPE stations. Cerebellum- 2MCQ		Histology of cerebellum 1OSPE station	Modulation of motor function by basal ganglia and cerebellum 4 MCQs	8 MCQ and 3 OSPE stations
Sensory modulation and homeostatic functions: Thalamus, hypothalamus and reticular formation.	Thalamus- 1 MCQ, 1 OSPE station	Development of Nervous system 4MCQ		Reticular activating system and consciousness- 2 MCQ Hypothalamic nuclei and functions-2MCQ	9 MCQs, 1 OSPE station
Complex brain functions: Sleep, emotions and memory	Limbic system- 1 MCQ, 1 OSPE station			Physiology of sleep- 2MCQ Epileptogenesis 2 MCQ EEG 1 OSPE station	5 MCQs and 2 OSPE stations

Themes/Disciplines	Anatomy	Embryology	Histology	Physiology	Total
Meninges and Ventricular system	Meninges and ventricular system 1 MCQ and 2 OSPE stations			Cerebrospinal fluid production and circulation 1 MCQs	2MCQs, 1 SAQ and 2 OSPE stations
Total	31 MCQ, 15 OSPE and 4SAQ	4 MCQ	3 OSPE	40 MCQ and 2 OSPE	75 MCQ, 4 SAQ, 20 OSPE station

3. Neuroanatomy: Loss of Acquired Knowledge

Many studies have documented that the study of Neuroanatomy has been marginalized in many medical school curricula [13, 19, 20], but little is known about how much of the neuroantomy is retained after graduation from medical school. A study done at the College of Medicine, University of Saskatchewan in 2008 indicated that graduating medical students do not retain enough knowledge of neuroanatomy to pass even the first year medical school neuroanatomy examination. They found that this loss of knowledge was far more significant than when retesting basic science knowledge for medical licencing examinations [21].

It is hardly surprising to find poor retention of basic science knowledge. Sinclair in 1965 found that retesting for the knowledge of lower limb anatomy, two years after the course; the results were not superior to random guessing [22]. Shulman in 1970 showed that the curves representing nonsense curriculum were similar to curves of forgetting knowledge gained in gross anatomy course by the end of clerkship years [23]. More recently, a study using the results of United States Medical licensing examination (Part- one) showed that there is a significant decline in the knowledge of basic science as students advance through years in medical school [21]. This shows that scoring high on a prior test does not necessarily relate with long term retention of essential knowledge [24]. There is a definite need to reemphasize basic science knowledge relevant to clinical encounters to medical students advancing through their curriculum, rather than teaching advanced level non-essential information [25]. Therefore, it is imperative for both faculty and students to make a distinction between the knowledge which is "worth remembering" and which is "non-essential". Given this premise, it is important for curriculum planners to emphasize clinically relevant neuroanatomy during the basic science course and revisit this core knowledge during the clerkship phase.

4. Future Directions in Neuroanatomy Curriculum

One of the major changes in the medical education in recent years has been a shift towards the outcome-based medical curriculum. We are now more interested in the question "What a clinical practitioner would be able to do" rather than "What should a medical student know". The learning outcomes for specific curricula and programs are built in broad outcome-based competencies. In the United States, there are six competencies which are required by the Accreditation Council for Graduate Medical Education for residents [26]. In Canada, the Royal College of Physicians and Surgeons of Surgeons established a structure of CanMED competencies [27]. This competency based frame work is now widely used for clinical training.

In putting together suitable learning outcomes for a neuroanatomy curriculum it is not only important to relate neuroanatomy with clinical practice, but also to include objectives which will integrate it with a wide range of competencies. An example of this could be that if students have difficulty in recalling neuroanatomy which they learned during basic science phase of the curriculum, they should regularly revisit information to the level necessary to understand clinical problems. This competency as described by ACGME is the willingness to assimilate and locate scientific knowledge relevant to patient care and utilize information technology in learning [26].

The Chair of the World Federation of Neurology research group on medical education argued in the British Medical Journal that emphasis in neuroscience teaching should first focus on patients' experience and clinical presentation and then on basic science explanation of disease [28]. Given this premise, we can make the teaching of neuroanatomy more effective and relevant, if the clinical presentation and treatment of the patient becomes foundational in future neuroanatomy curricula, rather than the knowledge about anatomical structures.

REFERENCES

[1] Hazelton L. Changing concepts of neuroanatomy teaching in medical education. *Teaching and learning in medicine.* 2011;23(4):359-64.

[2] Sabin F. Franklin Paine Mall. *The Story of a Mind.* 1st ed. Baltimore, MD: The Johns Hopkins Press; 1934. p. 342.

[3] Papa FJ, Harasym PH. Medical curriculum reform in North America, 1765 to the present: a cognitive science perspective. *Acad Med.* 1999;74(2):154-64.

[4] Ludmerer K. The university of Michigan Medical School: A tradition of leadership. In: Howell J, editor. Medical lives and scientific medicine at Michigan, 1891-1969. Ann Arbor, MI: The University of Michigan Press; 1993. p. 13-28.

[5] von Staden H. Herophilus: The Art of Medicine in early Alexandria. Translation and Essays.1st ed. Cambridge, UK: Cambridge University Press; 1989. p. 712.

[6] Swick HM. Medical professionalism and the clinical anatomist. *Clinical anatomy (New York, NY.* 2006;19(5):393-402.

[7] Mall F. On the teaching of anatomy. *Anat Rec.* 1908;2:313-35.

[8] Mall F. The anatomical course and laboratory of the Johns Hopkins University. *Bull Johns Hopkins Hosp.* 1896;7:85-100.

[9] Mall F. Liberty in medical education. *Phil Med J.* 1899;3:720-4.

[10] Pauly J, editor. The American Association of Anatomist,1888-1987:Essay on the history of Anatomy in America and a report on membership-Past and Present. Baltimore, MD: Williams & Wilkins; 1987. p. 292.

[11] Ludmerer K, editor. *Time to heal. American Medical Education from turn of the centuryto the ERa of Manged Care.* 1st ed. New York, NY: Oxford University Press; 2005. p. 514.

[12] Drake RL, McBride JM, Lachman N, Pawlina W. Medical education in the anatomical sciences: the winds of change continue to blow. *Anatomical sciences education.* 2009;2(6):253-9.

[13] Gartner LP. Anatomical sciences in the allopathic medical school curriculum in the United States between 1967-2001. *Clinical anatomy (New York, NY.* 2003;16(5):434-9.

[14] Heylings DJ. Anatomy 1999-2000: the curriculum, who teaches it and how? *Medical education.* 2002;36(8):702-10.

[15] Mateen FJ, D'Eon MF. Neuroanatomy: a single institution study of knowledge loss. *Medical teacher.* 2008;30(5):537-9.

[16] Crossman A, Neary D. Neuroanatomy: An illustrated colour text. London: Churchill Livingstone; 1998.

[17] Giffin BF, Drake RL. Gross anatomy of the head and neck and neuroscience in an integrated first-year medical school curriculum. *Anat Rec.* 2000;261(2):89-93.

[18] Cowan M, Arain NN, Assale TS, Assi AH, Albar RA, Ganguly PK. Student-centered integrated anatomy resource sessions at Alfaisal University. *Anatomical sciences education.* 2010;3(5):272-5.

[19] Haines DE, Hutchins JB, Lynch JC. Medical neurobiology: do we teach neurobiology in a format that is relevant to the clinical setting? *Anat Rec.* 2002;269(2):99-106.

[20] McKeown PP, Heylings DJ, Stevenson M, McKelvey KJ, Nixon JR, McCluskey DR. The impact of curricular change on medical students' knowledge of anatomy. *Medical education.* 2003;37(11):954-61.

[21] Swanson DB, Case SM, Luecht RM, Dillon GF. Retention of basic science information by fourth-year medical students. *Acad Med.* 1996;71(10 Suppl):S80-2.

[22] Blunt M, Blizar, PJ. Recall and retrieval of anatomical knowledge. *Br J Med Educ.* 1975;9:255-63.

[23] Shulman LS. Cognitive learning and the educational process. *Journal of medical education.* 1970;45(11):90-100.

[24] D'Eon MF. Knowledge loss of medical students on first year basic science courses at the University of Saskatchewan. *BMC medical education.* 2006;6:5.

[25] Gunderman R. Achieving excellence in medical education. London: Springer-Verlag; 2006.

[26] Accreditation Council for Graduate Medical Education. Common Program requirements: General competencies. [cited 2010 November, 10]; Available from: http://www.acgme.org/outcome/comp/GeneralCompetenciesStandards21307.pdf.

[27] Royal College of Physicians and Surgeons. CanMEDs. [cited 2010 November 10]; Available from: http://rcpsc.medical.org/publications/roles_e.html.

[28] Menken M. Demystifying neurology. *BMJ (Clinical research ed.* 2002;324 (7352):1469-70.

In: Education in Anatomical Sciences
Editor: Paul Ganguly
ISBN: 978-1-62618-488-6

Chapter 11

ISSUES CONCERNING ANATOMY IN PBL

Manoj Chakravarty*
College of Medicine, Department of Biomedical Sciences, King Faisal University, Al-Ahsa, Kingdom of Saudi Arabia

ABSTRACT

This chapter provides an insight into the teaching of anatomy from the perspectives of recent historical trends, examination of the status of anatomy in the Problem-Based Curriculum (PBL) and its comparison with traditional curricula.

Contextually, there is the interplay between the cognitive domain of the subject area, pedagogical methods, professional skills, students' approaches to learning, and the philosophy underlying the delivery of the curricular material along the PBL approach. The chapter deals with various issues, including the role of PBL, controversies regarding PBL curricular issues, and assessment strategies that could be considered important for an appropriate anatomy curriculum that is created to deliver emphasis on clinical application.

The chapter reinforces the role of anatomy in the PBL curriculum and suggests measures that need to be addressed to establish the role of anatomy as an indispensible and clinically relevant subject in PBL.

INTRODUCTION

Courses of study in human anatomy introduce students to terminology and structural organization used to describe the human body. Study of the human body forms the basic foundation for the professional practice of medicine.

Anatomy instruction should aim to provide an understanding of the dynamic three-dimensional organization of the body that is a pre-requisite for performing a relevant physical examination on any patient, understanding and demonstrating normal and abnormal physical

* E-mail: manoj.chakravarty@gmail.com.

signs, and interpreting results of imaging modalities that reveal internal structures of the body. In order to provide the basis for effective clinical reasoning, this understanding should contribute to knowledge of the body's systems and regions, including their development, function and pathophysiology.

Students' approaches to learning are associated not only with their conceptions of learning in general, but also specifically with their conceptions of what they are learning (Prosser and Trigwell 1999), in this case the learning of anatomy.

PBL

The ultimate goal of PBL is to educate students to be self-directed, independent, life-long learners. Through actively executing problem-solving processes and observing tutors' deal with problem-solving, reasoning, and meta-cognitive processes, PBL students learn how to think and learn independently. Problem-based learning is considered by many researchers to be the most innovative instructional method to date. Meta-analyses have indicated that PBL results in performances on clinical examinations are equal to if not better than performances following traditional teaching methods. (Norman, 1992), (Vernon, 1993), (Albanese, 1993), (Percac, 1998)

A primary assumption of PBL is that learning occurs when we attempt to solve problems. The most definitive finding regarding PBL is the superiority of PBL-trained stakeholders to be independent learners.

Problem-based learning is a methodology that offers an instructional solution to learning problems. The role of 'problems' in PBL is largely to serve as a stimulus or focus for students to direct their learning. The primary goal of PBL is to enhance learning by requiring learners to solve problems.

It is a pedagogical methodology that is problem-focused, student centered, self-directed, self-reflective and inclusive, where tutors are facilitators (not knowledge disseminators) who support and model the reasoning processes, facilitate group dynamics, probe students' learning, offer feedback and abstain from interjecting or providing direct answers to questions.

Problem based learning is an innovative and challenging approach to medical education – innovative because it is a new way of using clinical material to help students learn, and challenging because it requires the medical teacher to use facilitating and supporting skills rather than didactic, directive ones (Bligh, J. 1995)

PBL encourages the learner to work on problems for himself/herself. There is often initial surprise at not having to memorize tracts of isolated information and discomfort at the prospect of discovery rather than passive learning. The challenge of integrating experiences, previous knowledge, skills and activities with new knowledge and using this to solve clinically related problems, is stimulating for many students and is most appropriate in a community setting. (Glick, 1991)

PBL facilitates not only the acquisition and organization of knowledge but also the development of several other desirable attributes, such as communication skills, team work, problem-solving skill, self-directed learning, sharing information and identification of personal strengths and weaknesses (Neville, 1999).

Present Concerns with Teaching of Anatomy

The practice of Medicine relies on sound understanding of Anatomy. (Rudland, Rennie, 2003). Assessment options, comparison of teaching methods and faculty opinions must all form part of the decision-making process. Many consider anatomy teaching to be a casualty in modern medical education reforms. Indeed, senior doctors have commented on the lack of anatomical understanding in their trainees. (Hinduja, Samuel, Mitchell, 2005). The relevance of anatomy in the pre-clinical medical programme has often been brought into question. (Miles, 2005). As a result, the relative emphasis given to anatomy in the undergraduate curriculum has undoubtedly reduced. (Spielman, Oliver, 2005). This is a problem for those entering surgical specialties, where surgeons have commented that they are forced to take on the additional role of anatomy tutor for their trainees. Anatomy curriculum is undergoing international reformation but the current framework lacks uniformity among institutions. Optimal learning content can be categorized into the following modalities: (1) dissection/prosection, (2) interactive multimedia, (3) procedural anatomy, (4) surface and clinical anatomy, and (5) imaging.

The importance of multimodal teaching has been widely recognized and assessed. Nevertheless, there are still ongoing limitations in anatomy teaching. Substantial problems consist of diminished allotted dissection time and the number of qualified anatomy instructors, which will eventually deteriorate the quality of education (Sugand K. et al., 2010). In addition, Anatomy has suffered as a result of its failure to evolve and adapt quickly enough (Turley, 2007).

PBL and Clinical Anatomy

Prior to the introduction of PBL, establishing a full understanding of all normal anatomical structures was regarded as a necessary pre-requisite before the training in the clinical sciences. For nearly thirty years, there has been discussion of the decline in the undergraduate knowledge of anatomy amongst the surgical community. (Kaufman, 1997)

Thus, the anatomy curriculum requires regular updates in terms of presentation and content. However, there has been a lack of consensus regarding the way these modifications should be performed. In addition, adaptation of the curriculum has become problematic because the time, importance and academic credit allocated to anatomy has been progressively restricted to the benefit of other emerging disciplines, such as neurobiology or ethics. As a result, most of the departments constantly struggle to preserve essential—yet costly and time-consuming aspects of the curriculum such as dissections (Yiou, Goodenough, 2006)

Current PBL curricula offer a greater integrated approach such that the normal structural arrangement of the human body and organ-systems are presented and studied concurrently with abnormalities related to them. In the PBL curriculum, gross relationships are learned by students from prosected wet and plastinated specimens and models. These relationships are made more meaningful by the use of conventional radiographs, CT, and MRI images. (Reidenberg and Laitman, 2002).

Furthermore, surface anatomy and radiological anatomy are integrated into the Professional Skills program that runs concurrently with the weekly health-care problems on the various systems throughout the pre-clerkship phase of the curriculum (Ganguly et al., 2003).

The primary aim and focus of this program is for students to acquire relevant clinical skills and hands-on experience that helps them in clinical practice and diagnosis (Teichgraber et al., 1996; Boon et al., 2002; Tavares and Silva, 2002).

Other methods tried include the COATS program where the integration of clinical learning with conventional anatomy teaching aided the development of critical analysis skills and the understanding of the anatomical basis of clinical problems. This has been supplemented by case studies and presentations, radiological procedures and living anatomy (Issac, Chandi, Baidya, 1999). Medical students studying anatomy require a great deal of memorization to remember facts. Students complain that although they study anatomy thoroughly in their first year, they are unable to apply knowledge in the clinical year.

On the contrary, in the clinical year, they first realize that anatomical knowledge is extremely important as a basis for their knowledge of disease. (Pabst, 1993). PBL is effective at allowing students to ask questions about the purpose of their actions in anatomy and evokes curiosity beyond the set learning objectives (Slotnick and Hilton, 2006).

Role of IT in Anatomy in a PBL Setup

Computer aided learning has provided a major resource for the learning of anatomy with a gamut of software programs in almost all subject areas. Students are provided with study guides, interactive options, self-assessment questions to supplement existing methods of teaching and learning. This resource can additionally be of great assistance in enhancing problem-solving. Areas such as teaching through simulation, biostatistical applications such as SPSS, Computer assisted consultations, Computer assisted instructions and Medical informatics are all part of the expansive scope and application through the use of computers.

There is scope to include learning tools like magnified laparoscopic imaging sequences (Jiménez and Aguilar, 2009) and an accelerated video-recording of the entire alimentary canal using capsule endoscopy. Capsule endoscopy used in both normal and pathological cases led to half of medical students achieving a rate of 80% sensitivity in detecting small bowel lesions as novice readers (Chen et al., 2006).

As medical education moves towards problem solving and integration between different disciplines rather than by information acquisition alone, initiatives need to go beyond with the understanding and application of IT in the clinical decision making process and communication.

Assessment of Achievement in PBL

For the purpose of ensuring the domains of Validity, Reliability and Fidelity being mated to the constituent areas of anatomy through educational objectives, it has become necessary to evolve a cogent form of assessment that cuts across a broader spectrum of topics and provides a meaningful level of inter-disciplinary integration. Professional competencies in medicine are related to the coverage of cognitive, psychomotor and affective domains of knowledge, practical and clinical skills that are now being used more and more rather than tests for cognition alone as done in the past.

In this regard, anatomy examination questions are integrated, comprising of MCQ, Patient management problems, OSPE and internal assessment through the incorporation of outcomes (scores) of tutorial performance in PBL along with the use of standards for establishing minimal competence for judging minimal competence (Chakravarty et al., 2005).

In the practical setting for the assessment of interpretational skills and inter-disciplinary integration, different test items are constructed that include prosected cadaveric material, plastinated specimens, histological slides, imaging material, photomicrographs along with pathological specimens that adequately allows for in-depth application and coverage.

CONCLUSION

Learning anatomy involves focusing on the learning outcomes necessary if medical graduates are to make effective use of what they learned when confronted with real cases in professional practice (Bowden and Marton, 1998). However, the key to the reinvention of anatomy is the emergence of new entities such as 'clinical anatomy,' that introduce terms together with the foundation of organ structure and the general arrangement of organs into body systems, before launching into the detailed study of a specific system. Understanding gross anatomy along with its related areas of histology, embryology, radiology and being able to relate these to the pathophysiology and clinical features of a disease entity is essential within a PBL curriculum. This requirement is fulfilled through the active participation by faculty in curricular planning, design, implementation and evaluation. PBL curricula have the advantage of ensuring that students are immersed in a so-called spiral orientation towards learning, where basic concepts are introduced at an early stage of the undergraduate program, and these are sure visited at greater depth at regular intervals across the entire program. In reinventing itself via advancing, and not just retaining the best of the past and the present, anatomy has the capacity to lead by being an educational exemplar for other programs to reinvent themselves. This might even be of help to certain undergraduate medical courses facing the threat of being externally dismantled, as well as help to provide a secure foundation for future postgraduate training.

Problem-based learning is a total approach to education. PBL is both a curriculum and a process. The curriculum consists of carefully selected and designed problems that demand from the learner acquisition of critical knowledge, proficiency in problem solving, self-directed learning strategies, and team participation skills.

There is a great divergence in medical schools in teaching medicine in general and anatomy in particular (Older, 2004). It is well appreciated that anatomy should be learned as a dynamic subject for problem solving and application in the clinical practice towards delivery of quality health care (Miller et al., 2002). It is also recommended that the anatomy course should be spread over the entire undergraduate curriculum rather than delivered in a concentrated block in preclinical years, with the added benefit of introducing patient contact from the start of medical education (Sugand K, 2010) and this is very well suited both theoretically as well as practically to a PBL curriculum.

REFERENCES

Albanese MA, Mitchell S. 1993. Problem based learning: a review of literature on its outcomes and implementation issues. *Acad. Med.,* 68:52-81.

Bligh John, 1995. Problem based learning in medicine: an introduction. *Postgrad. Med. J.,* 1995; 71: 323-326.

Boon JM, Meiring JH, Richards PA. 2002. Clinical anatomy as the basis for clinical examination: development and evaluation of an Introduction to Clinical Examination in a problem-oriented medical curriculum. *Clin. Anat.,* 15:45-50.

Bowden J, Marton F. 1998. The University of learning: Beyond quality and competence in London:Kogan Page.

Chakravarty M., Latif N.A., Abu-hijleh, M., Osman, M., Dharap, A.S., Ganguly, P.K. 2005. Assessment of Anatomy in a Problem-Based Medical Curriculum. *Medical Education,* 18:131-136.

Chen GC, Enayati P, Tran T, Lee-Henderson M, Quan C, Dulai G, Arnott I, Sul J, Jutabha R. 2006. Sensitivity and inter-observer variability for capsule endoscopy image analysis in a cohort of novice readers. *World J. Gastroenterol.,* 12:1249–1254.

Ganguly PK,Chakravarty M, Latif NA, Osman M, Abu-Hijleh M. 2003. Teaching of Anatomy in a Problem-Based Curriculum at the Arabian Gulf University: The New face of the Museum. *Clinical Anatomy*; 16:256-261.

Glick SM. Problem-based learning and community oriented medical education. *Med. Educ.,* 1991; 25: 542-5.

Hinduja, K, Samuel R, Mitchell S. 2005. Problem-based learning: Is anatomy a casualty? *Surgeon,* 3: 2; 84-87.

Issac, B., Chandi, G., Baidya, R. 1999. Innovations in teaching anatomy. *Medical Education,* 33:628-29.

Jime´nez AM, Aguilar JF. 2009. Laparoscopy: Learning a new surgical anatomy? *Anat. Sci. Educ.,* 2:81–83.

Kaufman, MH. Anatomy training for surgeons- a personal viewpoint. *J. R. Coll. Surg. Edinb.,* 1997; 42:215-6.

Miles KA. 2005. Diagnostic imaging in undergraduate medical education: An expanding role. *Clin. Radiol.,* 60:742–745.

Miller S, Perrotti W, Silverthorn D, Dalley A, Rarey K. 2002. From college to clinic: Reason over memorization is a key for understanding anatomy. *The Anatomical Record;* 269: 69-80.

Neville AJ. 1999. The problem-based learning tutor: Teacher, Facilitator, Evaluator, *Medical Teacher.,* 21: 393-401.

Norman GR, Schmidt HG. 1992. The psychological basis of problem-based learning: a review of the evidence. *Acad. Med.,* 67:557-565.

Older J. 2004. Anatomy: A must for teaching the next generation. *J. R. Coll. Surg. Edinb. Irel.,* 2 (2):79-90.

Pabst, R. 1993. Gross Anatomy: An outdated subject or an essential part of a modern medical curriculum? Results of a questionnaire circulated to final year medical students. *The Anatomical Record*; 237: 431-433.

Percac S, Armstrong EG. 1998. Introducing a problem-based learning anatomy course in a traditional curriculum: a Croatian experience. *Med. Teacher*, 20:114-117.

Prosser M, Trigwell K. 1999. Understanding learning and teaching: The experience in higher education. Buckingham: SRHE and Open University Press.

Reidenberg JS, Laitman JT. 2002. The new face of gross anatomy. *Anat. Rec.,* 269:81-88.

Rudland JR, Rennie SC. 2003. The determination of the relevance of basic sciences learning objectives to clinical practice using a questionnaire survey. *Med. Educ*., 37:962–5.

Slotnick HB, Hilton SR. 2006. Proto-professionalism and the dissecting laboratory. *Clin. Anat.,* 19:429–436.

Spielmann PM, Oliver CW. 2005. The carpal bones: a basic test of medical students' and junior doctors' knowledge of anatomy. *Surgeon,* 3:257–9.

Sugand K., Abrahams P., Khurana, A. 2010. The Anatomy of Anatomy: A Review for its Modernization. *Anat. Sci. Educ*., 3:83–93 (2010).

Tavares MAF, Silva MC. 2002. Evaluation of the Clinical Anatomy Program in the Medical School of Porto by two cohorts of students. *Clin. Anat*., 15:56–61.

Teichgraber UKM, Meyer JMA, Poulsen Nautrup C, von Rautenfeld DB. 1996. Ultrasound anatomy: a practical teaching system in human gross anatomy. *Med. Educ*., 30:296-298.

Turley BW. 2007. Anatomy in a modern curriculum. *Ann. R. Coll. Surg. Engl.,* 89:104-107.

Vernon DTA, Blake RL. 1993. Does problem-based learning work? A Meta analysis of evaluative research. *Acad. Med*., 68:550-65.

Yiou R., Goodenough D. 2006. Applying problem-based learning to the teaching of anatomy: the example of Harvard Medical School. *Surg. Radiol. Anat.,* 28:189-194.

In: Education in Anatomical Sciences
Editor: Paul Ganguly

ISBN: 978-1-62618-488-6

Chapter 12

INNOVATION IN ANATOMY TEACHING

Claudia M. Diaz
RMIT University, Bundoora,
Melbourne, VIC, Australia

ABSTRACT

Effective anatomy teaching and learning involves student engagement through a passionate teaching style and creative, purposeful learning activities that align with the professional goals of students. Traditionally students find anatomy difficult as it involves many new concepts and complex terminologies that students find dull and labor-intensive and they often concentrate their efforts on “memorizing” lists of names, a time-consuming “surface” approach. Furthermore, many universities now offer fewer contact hours to teach anatomy and often there is also a limited availability of cadavers for study. The challenges of teaching anatomy in the modern world require anatomy to be taught more effectively, in less time and often with reduced resources. These modern challenges provide an exciting opportunity for the development of innovative approaches for teaching and learning anatomy that move away from the traditional didactic methods towards more engaging and stimulating approaches using practical, hands-on methods that have clinical significance and are related to the professional competencies required. Passionate (stimulating) learning approaches can lead to deep learning through understanding the information and fixing it into long-term memory and the ability to place this information into a broad picture relevant to their future professions. The specific approaches that have been described are “'whiteboarding” and drawing, use of play-doh, movement/singing/dancing, surface anatomy, and body painting which all encourage multi-sensory, 'hands-on' learning. Taking these new perspectives on traditional methods of teaching anatomy provides a much more engaging, motivating, inspiring and enjoyable environment for student learning. These new teaching and learning approaches produce students that can achieve the desired learning outcomes (higher grades), and are engaged, confident and self-direct learners.

Keywords: Anatomy, student, engagement, experiential, self-directed, didactic, learning

1. Intended Outcomes of This Chapter

1. To understand the importance of engagement to student learning.
2. To understand the challenges involved in teaching and learning anatomy.
3. To identify the different models for teaching and learning anatomy.
4. Beyond the teaching – a focus on learning! To identify the need for innovative approaches to improve achievement of learning outcomes.
5. Learning outcomes.
6. Conclusion.

2. Introduction - Engagement: The Key to Successful Learning

Effective student engagement is a result of both an interested, passionate (stimulated) teaching style [1] and students carrying out meaningful learning activities through interaction with others and worthwhile tasks [2-4]. Consistent with engagement theory, these meaningful learning activities must make learning a creative, purposeful activity, they should be self-directed and experiential (project-based) and aligned with students' professional or future career goals [2-4]. This alignment with the students' future career interests increases student motivation and satisfaction as students are driven to conduct their own projects rather than answering sterile textbook problems.

Interactions with peers are also valuable for developing planning, management and social skills needed for collaborative learning. The process of collaboration forces students to clarify and articulate their problems that in turn facilitates solutions. Collaboration is also an important factor in increasing motivation of students to learn which is particularly important with first year university students with high dropout rates [5]. The first year of university is the ideal time to maximize student engagement and in no subject more so than Anatomy.

3. Challenges in Teaching Anatomy

Teaching Anatomy faces inherent and contextual challenges and learning anatomy has always been a challenge for medical and health science students. Anatomy is traditionally a subject that is difficult for first year University students to learn as it involves many new concepts and complex terminologies that students find dull and labor-intensive [6]. Students often concentrate their efforts on "memorizing" lists of names, a time-consuming "surface" approach. Furthermore, with the changing face of medical and health education all around the world many universities now offer fewer contact hours to teach anatomy [7-13] and often there is also a limited availability of cadavers for study. Anatomy teaching at some Australasian medical schools have been slashed by eighty percent, with more than ninety percent of all teaching occurring in the early years [10, 14]. Although there may be some benefit in integrating anatomy teaching into the later years of undergraduate programs, this is not common. There have also been reports that a lack of anatomical knowledge has led to an increase in medico legal claims [15, 16]. In addition, classes are large and extremely diverse with respect to students' entry level, prior experience, scientific literacy levels, cultural backgrounds and professional fields. So, the challenges of teaching and learning anatomy in the modern world require anatomy to be taught more effectively, in less time and often with reduced resources. This provides an exciting time and opportunity for the development of innovative approaches to teaching and learning anatomy [17-19]. The use of new technologies and teaching methods should provide a more learner centered approach with better teaching, learning and understanding for continued learning.

Although there have been notable cuts to anatomy teaching, students still acknowledge the importance of anatomy to their professions and have created avenues to increase anatomy education [20, 21]. In a rural Australian medical school, medical students formed an Anatomy Society that has established a student-initiated extra-curricular anatomy education program. The program supplements anatomy teaching by providing clinically oriented, cadaver and image-based anatomy education workshops organized and run by students [20, 21].

4. Models of Teaching and Learning Anatomy

Anatomy has traditionally been taught using didactic approaches including the didactic lecture and practical class setting that encourage the memorization of concepts and names and learning from two-dimensional diagrams [22]. Research suggests that didactic lecture/presentation styles do not encourage student engagement. The didactic approach proves very difficult for students with poor spatial abilities as they struggle to form three-dimensional (3D) relationships from the two-dimensional diagrams [22, 23]. It has been recommended that different teaching techniques be implemented in the curriculum to accommodate students with lower spatial abilities, including dissection/prosections and videos of three-dimensional anatomical structures in an anatomy setting, 3D models and several 3D videos [22]. Furthermore, the didactic approach does not encourage collaboration, development of problem-solving skills [23] or teach students self-directed or life-long learning skills that are important in the development of professional competencies. The use of didactic methods has resulted from the belief that anatomy is largely "content-driven" and not "skills-based" [24]. Didactic teaching tends to involve surface learning rather than deep learning and involves a low level of cognitive engagement whereby students tend to use rote learning and memorization to recall facts [25]. Strategic learning is "deliberate surface learning" where students only learn what they need for the examination, learning the minimum to get by [25]. Strategic learning is a conscious, strategic move to obtain marks or credit and strategic learners make an informed choice about when to be a deep learner and when to be a superficial learner. Surveys of medical and health science students have reported bad scores for didactic teaching compared with practical methods of teaching anatomy [24]. Furthermore, a more recent study has also reported quantitative and qualitative data supporting the preference by medical and health science students for practical, hands-on methods compared with didactic ones [19].

Anatomy is typically taught using didactic lectures in a lecture theatre, followed by a practical class in the anatomy laboratory. Some Anatomy laboratories are fortunate to have access to cadaveric materials for students to study in addition to further resources such as skeletons, isolated bones and plastic models. Other anatomy laboratories may offer additional resources such as x-rays, plastinated specimens, computer programs, surface anatomy and body painting. Typically an anatomy practical class in the anatomy laboratory is run using a working guide or workbook for students to read and follow throughout the class. There are often academics or tutors available to "teach" students (using a didactic, surface approach) in these classes.

For anatomy teaching to be effective, it needs to have clinical significance and be related to the professional competencies required by new graduates, as students need to understand its relevance to their future professions. Assessment also needs to focus on integration and application of anatomical knowledge where possible.

Anatomy teaching will best meet the learning outcomes and professional competencies required by graduates if it supports the model of experiential learning which is the process of making meaning from direct experience [26, 27]. Medical and health science students have been shown to have 'direct encounters with the phenomena being studied', and 'acquire and apply knowledge, skills and feelings in an immediate and relevant setting' using innovative anatomy teaching approaches [19]. In this type of learning the students participate in the four stages of experiential learning: (i) they carry out particular actions in the anatomy laboratory as part of these teaching approaches, (ii) they learn and reflect about these effects in particular instances (they learn by 'doing'), (iii) they come to understand the general principles (they start to experience deeper learning) and (iv) they learn to apply through action in new circumstances (they are able to apply these techniques to new areas encountered). Effective anatomy teaching and learning requires a high level of cognitive engagement, leading to a deep approach to learning where students make meaning of the content and can apply it to other contexts [28]. This is opposed to the surface approach where students think it necessary to remember the knowledge using rote learning.

The principal factors underpinning successful learning are the desire to learn (motivation, interest, enthusiasm), needing (necessity, surviving), doing (practice, repetition, experience), feedback (reactions, results) and digesting (making sense of it all) [25].

5. Beyond the Teaching – A Focus on Learning!

Course aims and learning outcomes should be determined before appropriate approaches for teaching anatomy can be chosen. Students first need to want to learn and then they need to follow a step-by-step path that will lead them to achievement of the learning outcomes. Wanting to learn is important as it involves emotions and feelings and is a powerful factor in helping learning to happen [25]. Effective anatomy teaching therefore requires linear mapping, that is achievement of course aims and learning outcomes through following a sequence of learning steps and experiences that will take students on a path to achievement. Well-designed learning outcomes help students identify their targets and work systematically towards demonstrating achievement of these targets [25]. Learning outcomes are important in providing students with an indication of the level and depth of learning required.

It has recently been postulated that multidisciplinary first year students can be stimulated to learn anatomy for life via innovative, pro-active approaches to improve engagement and learning outcomes [18, 19]. These innovative teaching and learning approaches strive to provide a nurturing and motivating environment for students that will make learning both enjoyable and stimulating. The use of these new approaches provides the stepping-stones for achieving the necessary linear mapping for effective anatomy learning. This study suggests that an enthusiastic and passionate teacher is important in producing enthusiastic student learners. Implementation of different, more stimulating and engaging teaching approaches is important in changing the historical views regarding the teaching and learning of anatomy. By making anatomy engaging, stimulating and fun, and by helping students to be pro-active learners, the foundations are being laid for a new approach to learning anatomy [18, 19]. The use of passionate (stimulating) learning approaches can lead to deep learning through understanding the information and fixing it into long-term memory and the ability to place this information into a big, broad picture relevant to their future professions.

Fundamental to these new approaches are several key principles. First, the use of human cadaveric tissues is central to students' learning experience of anatomy. The invaluable dissections and prosections are, and will always be the primary focus of engaging practical classes. Second, new approaches for teaching anatomy should be based on "engagement theory", that is students need to be engaged in meaningful learning activities through interaction with others and worthwhile tasks [29, 30]. Consistent with Kift [31], students need to be engaged, supported and 'fit in' in their first year of university, in fact from day one. These new approaches are consistent with the view that interested and passionate teachers are critical to student engagement [1]. Furthermore, good anatomy teaching must embrace a philosophy of inclusion of all students from diverse backgrounds, with different abilities and challenges. At most universities it is common for the best lecturers to teach the upper years of degree programs, however, if students are to be engaged and stimulated and taught life-learning skills in their first year of University, it would be ideal to have the best lecturers teaching first year.

Our role as anatomy teachers is *not* to teach students anatomical facts, but rather to teach them the learning skills that will serve them throughout their student and professional lives. In this way students can be taken from the traditional view of anatomy as a subject that requires surface learning (rote learning, memorisation) to one that can lead to deep learning through understanding and the ability to place information into a broad, big picture that also makes it relevant to the students' realities.

Accordingly, ranges of innovative, pro-active approaches that complement the use of prosected tissues have recently been trialled [18, 19]. These techniques are based on an old-fashioned, simple, "hands-on" approach where classes are *not* didactic and they aim to achieve maximum student engagement. As a result of these approaches, health science students have been shown to significantly increase their engagement in the anatomy course, improve their cognitive learning skills, become more self-directed in their learning, and have a high level of (long-term) anatomy knowledge. Therefore, while the principal goal is to support and assist first year students develop their knowledge and learning skills in anatomy, these innovative strategies can also produce *self-directed, life-long learners.* Data suggests that year one health science students using these new approaches have a high level of engagement in the course (via a 2010 and 2011 survey that compares health science students to Medicine students who undertake a similar anatomy course that has a far more didactic approach). Also, there is evidence that the health sciences students are confident in being able to remember the anatomy curriculum and can summarize and synthesize concepts and facts to make them easier to learn – not only for anatomy, but also for other subjects [18, 19].

In keeping with this reconstructed approach to teaching anatomy the laboratory becomes the key focus for action where team activities encourage 'hands-on', multi-sensory learning by all students. To ensure consistency and a team approach it is also important to have a team of well-trained tutors who have a significant impact in fostering small group team work in the laboratory environment, further encouraging students to develop self-directed learning skills. Tutors should be trained to encourage maximum student engagement using the prosections and the innovative teaching approaches. It has been reported that using these new approaches with tutors also results in a higher retention rates as tutors appear to be engaged themselves, happier and more motivated to improve their own anatomy learning [18, 19]. Human cadaveric tissues and a team of well-trained tutors are central to the learning experience of anatomy students and ensure that we achieve the priority health concerns of the communities that our students have a mandate to serve. However, we must also consider the option of using these innovative techniques on their own at universities where cadaveric tissues are not available for study.

The specific approaches that have been described are "'whiteboarding" and drawing, use of play-doh, movement/singing/dancing, surface anatomy, and body painting that all encourage multi-sensory, 'hands-on' learning. Staff encourage the creativity and enthusiasm of students, by actively modelling these approaches. They also engage students in an exploratory way, analogous to the process of dissection, and are therefore invaluable in programs that offer dissection or prosections and in programs where cadaveric tissue is not available. Using these approaches, the laboratory is ideally arranged so that students work in groups around a dissection table with a whiteboard at one end and they are required to work together using these approaches to demonstrate that they have a deeper understanding of the structures and relationships before or while they study the prosections. Students are not provided with extensive workbooks, rather they are given brief guidelines with a list of learning objectives, thus placing the onus on the students to prepare appropriately for each practical class (self-directed, deep learning).

Techniques such as functional anatomy and body painting have been described previously [32-36], however, no other study has used all these approaches together, at the same time within each anatomy practical class. Furthermore, these "old-fashioned" approaches do not form a standard part of anatomy curricula in Australia and so in that way are innovative approaches for teaching anatomy in Australia. Many medical and health science programs have focussed on developing modern anatomy curricula based on technological advances such as DVDs, computer programs and three dimensional modelling [23], however few have focussed on more traditional hands-on approaches.

These practically based teaching and learning approaches appear to be essential for skill acquisition. Studies to determine student attitudes to anatomy have reported that practical methods of teaching anatomy scored more highly than didactic teaching, and there was greater "fitness for purpose" for teaching using cadaveric material and for living and radiological anatomy [18, 19, 24].

A. Use of cadavers/prosections. The use of dissections and prosections is central to learning anatomy and the other approaches discussed below are ideally used in conjunction with the study of cadaveric tissues. Studying anatomy using cadavers (dissections) and prosections helps students to understand the 3D structure of the human body through self-discovery, exploration and observations. It also helps to develop spatial reasoning skills to understand and interpret imaging data [37] and it helps with visualization of three-dimensional spatial relationships between structures, and it helps to understand inter-relationships of anatomical structures. At present, most anatomy laboratories use prosected cadaveric tissues for study (Figure 1).

The use of prosections maximizes the scarce resource of cadavers making more time available for teaching, learning and revision [37]. Amongst professional anatomists dissection is still the preferred mode of teaching anatomy [38]. Cadaveric dissection has been reported to be the most "fit for purpose" as it provides appreciation of the three-dimensionality of the body and with strategies for experiential learning, followed by prosection and demonstration and by living and radiological anatomy [38]. The use of cadaver dissection has been reported to encourage collaborative learning, but did not improve knowledge acquisition [39]. Medical students would prefer that anatomy be taught practically (via dissection, use of prosections, with living and radiological anatomy) than theoretically (via didactic teaching, models) [24].

Furthermore, many universities are also using plastinated specimens, cadaveric materials that are preserved using plastics [40]. Plastinated specimens are safe to handle and durable, they can be handled repeatedly by students and do not require special storage, however the plastination process also alters the characteristics of the tissue more than embalming and so makes the tissues rigid to handle. While dissections and prosections are invaluable resources for learning anatomy, they do not always provide a realistic or accurate view of the living body.

Although dissection is for many the preferred method for teaching anatomy, the process of dissection is slow and not compatible with the current issue of reduced hours to teach anatomy at many universities.

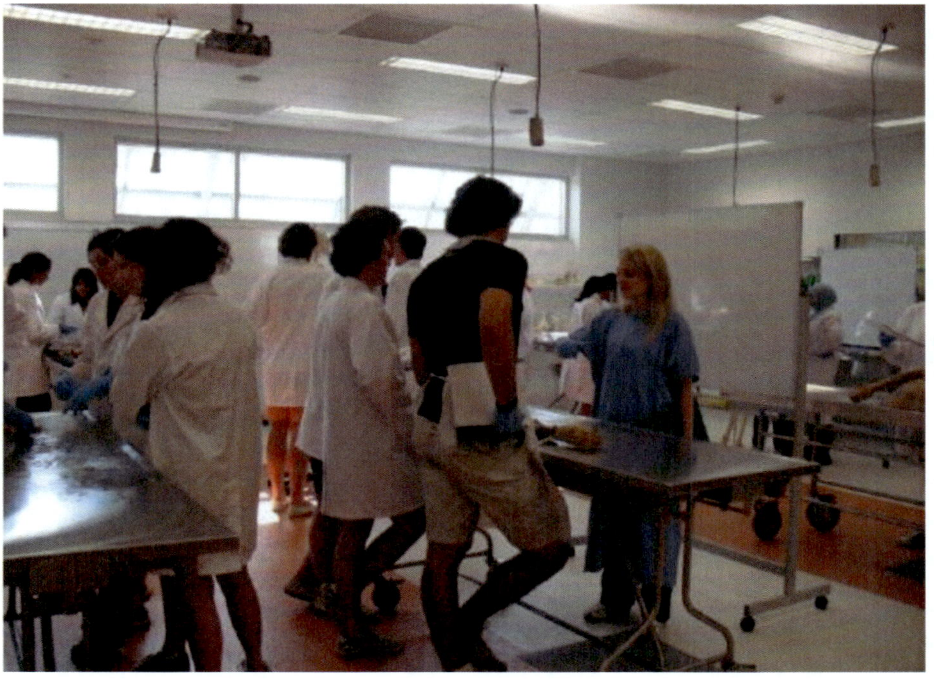

Figure 1. This is a typical practical class with first year anatomy students using prosections to complete anatomy practical class guidelines.

The exploratory behaviour exhibited by students during dissection can be replaced by the exploratory behaviour needed for the innovative methods described below. In fact these innovative approaches may become the primary teaching technologies to teach anatomy at universities that do not have access to cadaveric tissues. The following teaching and learning approaches are dependent on the initial visualization of the learning problem by students, that is, they first need to observe and try to understand what they are about to learn. This will often involve the visualization of two-dimensional images. The innovative approaches will then take them on a path of discovery, that will be a stimulating and motivating experience and will help them to start to recreate structures into a three-dimensional portrayal.

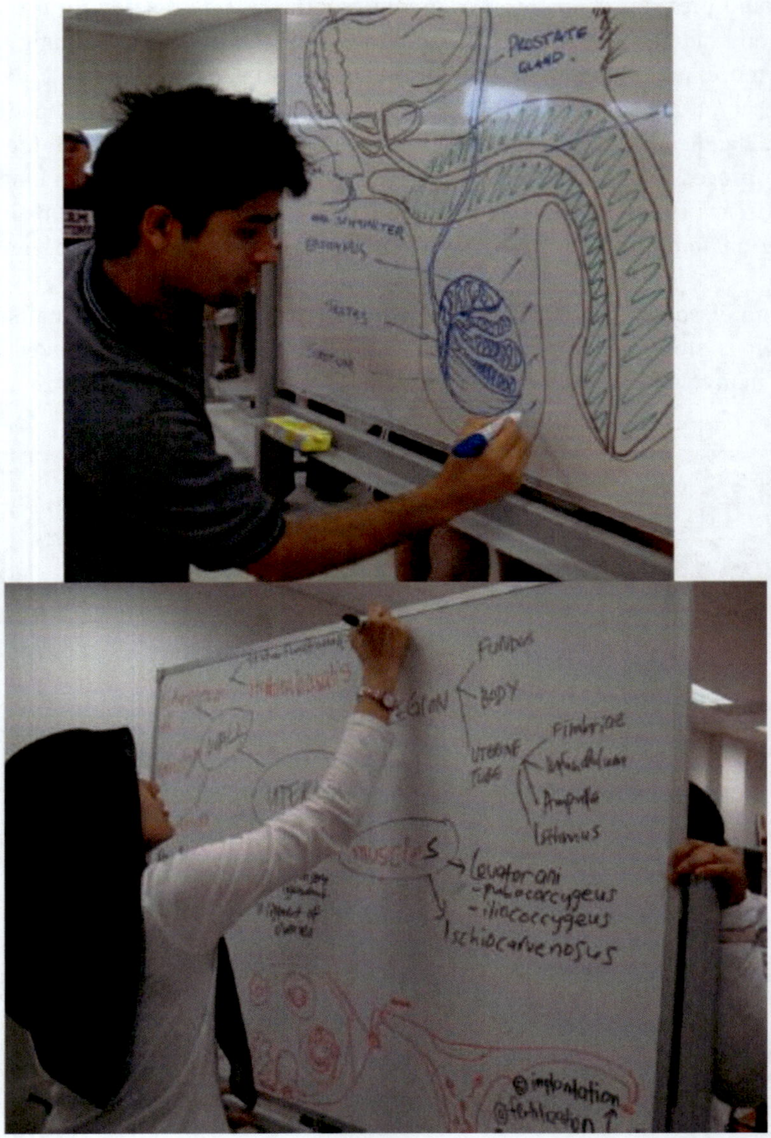

Figure 2. In practical classes, first year anatomy students use "whiteboarding" activities as part of their anatomy learning activities.

B. White-boarding and drawing. (Figure 2) The use of a whiteboard and drawing are learning approaches that appeal to visual learners. Students learn to use a whiteboard as a learning tool to summarise and synthesize concepts and facts. The whiteboard is a powerful learning tool that acts as a clinical communication tool by using diagrammatic relationships in preference to written or verbal descriptions. Typically students use schematic diagrams, flowcharts and simple drawings to learn anatomy in practical classes. Whiteboarding allows students to manipulate a complex set of relationships in a diagram, thus understanding the relationships, remembering them and being able to analyse their parts [41]. This promotes a "deep" approach to learning [42]. Whiteboarding supports the dual code theory that both visual and verbal information is used to represent information, in this way forming mental images that aid learning [43].

Furthermore, whiteboarding develops the advantageous skill of mind mapping that requires active engagement that leads to greater learning [41]. Mind mapping consists of visual, non-linear representations of ideas and their relationships [44] and involves connected and related concepts that require spontaneity and creativity. The use of colours, line thickness, symbols, codes, emphasis, pictures and diagrams are all strategies that help with knowledge recollection. Mind mapping has been reported to assist with memory retention as pictures and structured diagrams are easier to remember than just words and help with understanding of difficult topics [41].

The advantages of mind mapping are the unconstrained nature and there are no limits on the ideas and links that can be made, thus promoting "brainstorming" [41]. Furthermore, it has been reported that whiteboarding is an advantageous skill for students as it is applicable to other subjects other than anatomy. Indeed the author has had many students become so enthusiastic about whiteboarding that they have purchased their own whiteboards for home and used them for anatomy and other subjects.

Similarly, visual learners and artists benefit and are encouraged to draw anatomical structures on paper or similar. Many anatomists still use drawing on a blackboard, whiteboard or multimedia display to allow students to develop the skill of building a two-dimensional into a three-dimensional organization [35]. Drawing is a great way to develop spatial skills and it is easy to challenge students to do so [22].

Learning by drawing surface anatomy of abdominal organs has been shown to improve grades [34] and anatomy drawing has also been shown to improve grades for first year physiotherapy students who described the experience as enjoyable and extremely valuable [45]. Within any cohort of students there are always a few students who are exceptional artists and their abilities and enthusiasm should be encouraged and shared with the class to further anatomy learning for all (Figure 3).

C. Use of Play-doh (Figure 4) The use of play doh is an approach that appeals to tactile learners. Students build anatomical structures using play doh. This approach is enjoyable and reinforces learning while assisting students to develop a real 3-D image of structures and their relationships. Furthermore, students can develop their imaging skills by sectioning their play doh creations; this is particularly useful when studying the brain for example, as students can build a whole brain, then section it both sagittally and coronally for further study. Visual and tactile learners also benefit from handling bones, skeletons, and plastic models in anatomy practical classes.

D. Movement/singing/dancing (Figure 5) The use of movement, singing and dancing are learning approaches that appeal to students who learn by 'doing'. Performing body movements with weights, a hula hoop, etc., are encouraged in many anatomy practical sessions, as well as singing and dancing (where appropriate), as these involve movement and this 'doing' appears to assist in long term retention of information for many students.

An example of such an activity is the author who as part of the reproductive anatomy lecture every year re-enacts a life-size living uterus, and even throws a blanket over herself to represent the broad ligament. Students have reported that this is something they will never forget.

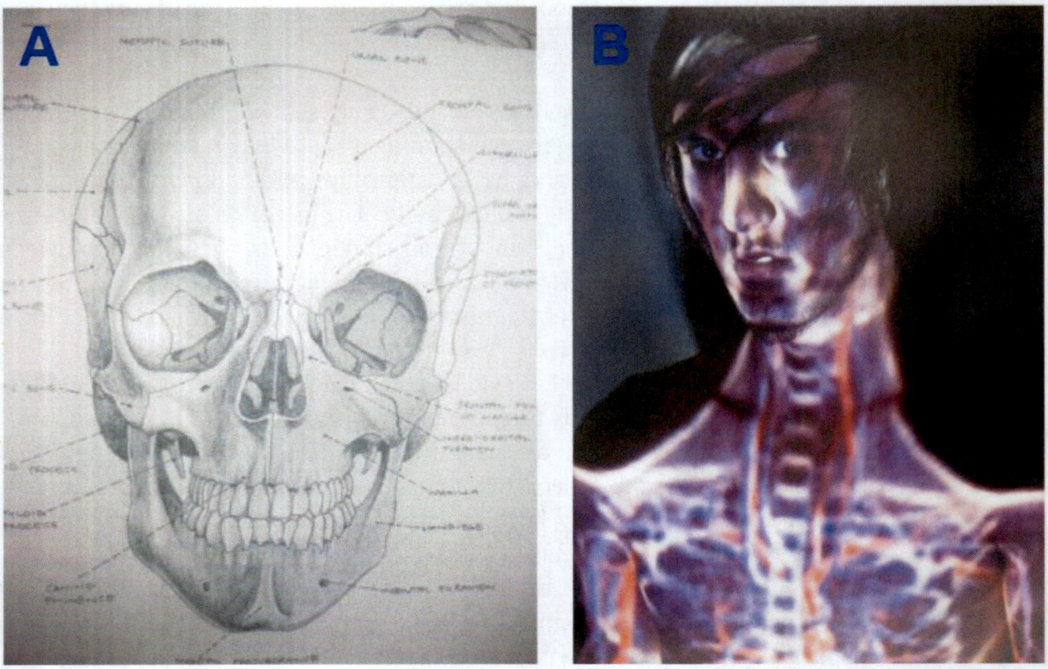

Figure 3. In anatomy practical classes first year anatomy students often use art as a learning approach as seen in this drawing (A) and visual arts picture (B).

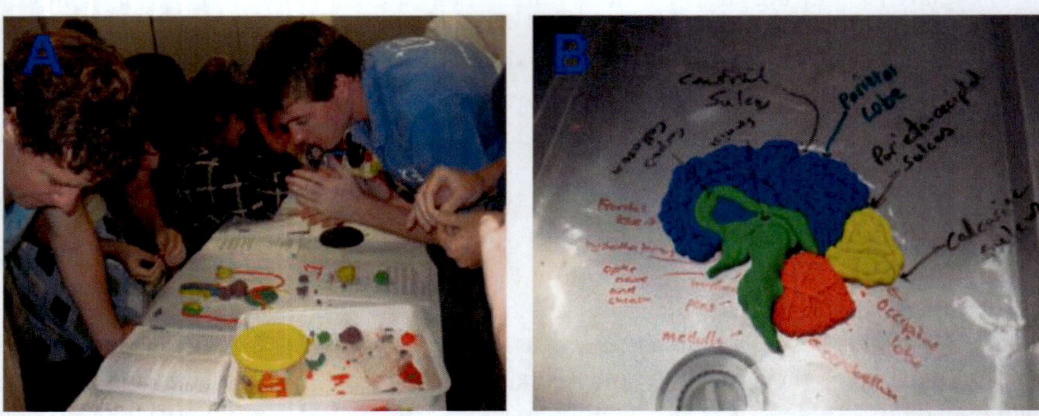

Figure 4. In anatomy practical classes first year anatomy students build anatomical structures using play doh.

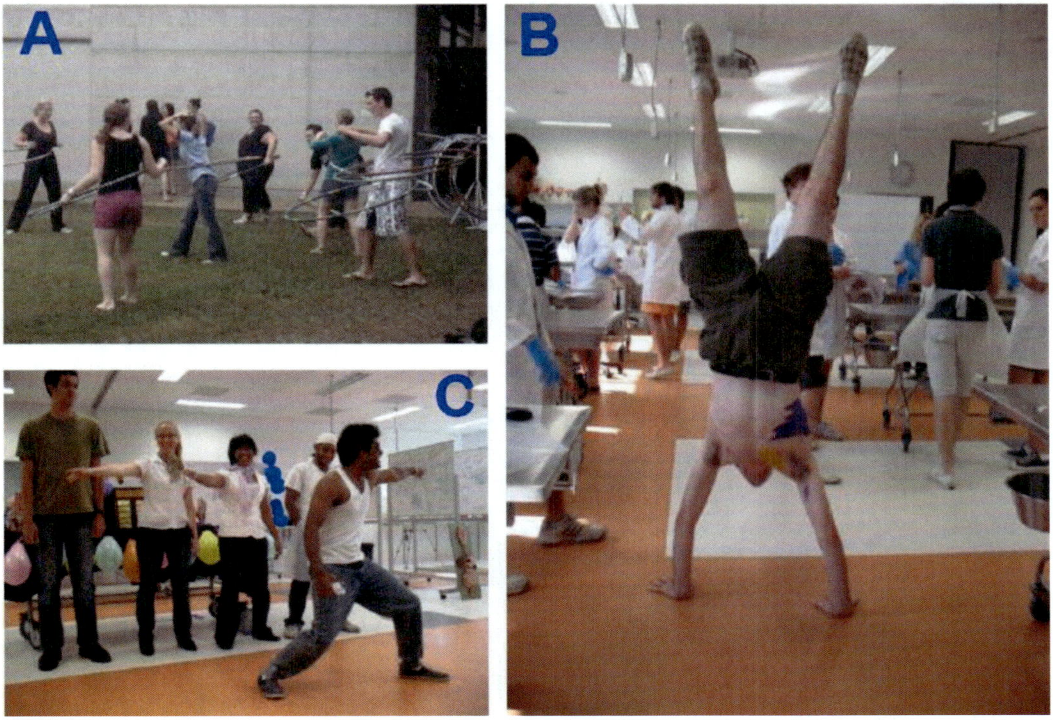

Figure 5. Popular learning approaches in anatomy practical classes can include movement, singing and dancing. A. Hula hooping lessons outside the lecture theatre when studying abdominal muscles. B. A student performing functional anatomy when studying pectoral girdle muscles. C. Students performing a song they prepared about the gastrointestinaltract.

E. Surface Anatomy/body painting (Figure 6-8) Using learning approaches such as surface anatomy, including body painting, students consolidate what they learn with prosections by looking at the relevant surface anatomy painted on a live "Anatomical Man/Woman" (themselves, classmates, family or friends). The use of living bodies to study surface anatomy has been neglected [37]. The study of surface anatomy is particularly important in studying the musculoskeletal system as students can see structures moving and functioning. Surface anatomy also teaches students to observe, examine and interact with a living person [37]. Body painting has become a popular surface anatomy technique as it is a very engaging way for students to learn anatomy by looking, identifying surface landmarks, palpating, drawing, and then painting. This application to live bodies is what they will need to do in their professional lives when working with patients or clients. The use of body painting to teach anatomy has been reported as an important adjunct to surface anatomy and clinical skills [18, 33, 36]. Diaz [18] has reported the use of body painting as a powerful learning approach with medical and health science students. In this study, a fully body painted model was brought into the first health sciences musculoskeletal anatomy practical class of the year to surprise, inspire and motivate students as they witnessed how engaging and beautiful anatomical body painting can be. Student participation in body painting increased in subsequent classes, as they became very confident and motivated to carry out this learning approach as either models or painters.

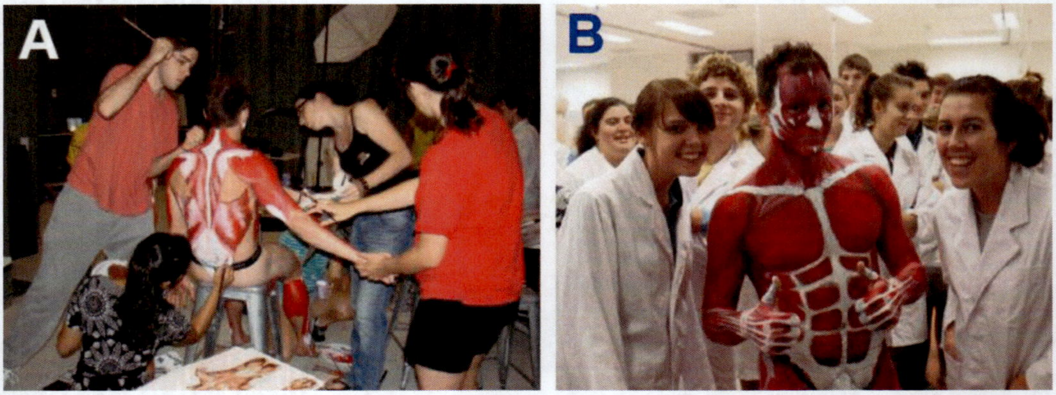

Figure 6. Body painting projects are conducted several times a year to inspire and motivate student participation. A. A student model is fully body painted by student artists; B. the fully painted model is brought into the first musculoskeletal class of the year to surprise and inspire students.

Figure 7. Surface anatomy activities, including body painting are integral parts of most anatomy practical classes.

Emphasis is placed on the process of body painting, that is the identification, palpation, drawing in of landmarks, drawing in of origins and insertions, and finally the painting (Figure 7). In this way all students can participate and gain the same benefit without having to be great artists.

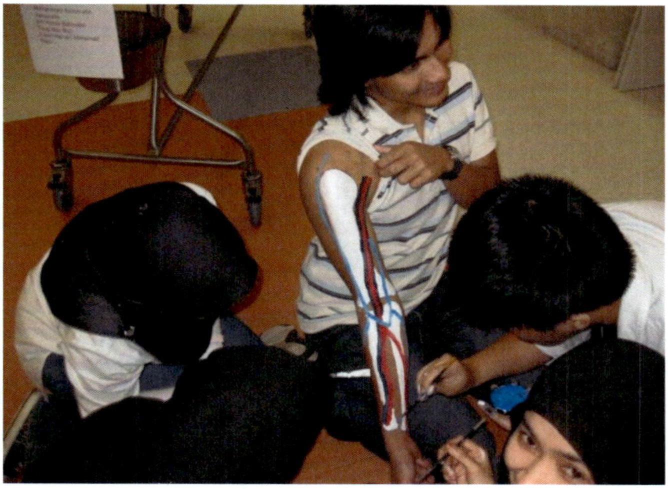

Figure 8. First year Muslim medical students participating in body painting activities in their anatomy practical class.

Body painting is an approach that suits all students but cultural sensitivity and careful negotiation are necessary to ensure that students are comfortable carrying out the activities. For example at a regional Australian medical school Muslim students were very enthusiastic about participating in body painting activities and through careful negotiation they were able to find an area of the body that they were comfortable to body paint, and indeed they came second in a body painting competition (Figure 8).

Students have also reported that they participate in body painting activities in their free time with classmates and friends, further supporting this as an excellent and engaging learning approach.

F. Anatomy and technology. These hand-on learning activities are invaluable to students in their first year of anatomy. However, in the modern world of constant change and technology, anatomy must also embrace new technologies and incorporate them into our teaching practices as complementary resources to assist student learning. Many advances have been made in the area of anatomy computer and three-dimensional resources such as Anatomedia, a computer program that studies the human body from four perspectives: regions, systems, dissection and imaging [46, 47]. The modules examined are general anatomy, back, thorax, abdomen, pelvis, upper limb, lower limb, head and neck [47]. The visible human project also presents 1871 cross sections of a male, sliced every 1mm, and is an excellent and economical resource for students for self-directed study. Reports have indicated that there is a high demand for online anatomy resources as students report them to be more motivating and fun than textbooks [48]. Furthermore, Hallgren et al. [49] reported that students who used web-based material to learn anatomical landmarks compared with a control group that did not access to the material obtained better examination results. Computer-based 3D technology is being used internationally to teach anatomy and has been reported to increase anatomical knowledge and recall [23, 50]. Computer technology helps with information mapping needs [41] and offers alternative teaching methods to a cadaver-based course [51]. Wright [51] described the use of three-dimensional anatomical virtual dissection software, anatomical models coupled with a learning management system to offer web-based learning. Kotzé et al. [52] reported the beneficial use of full body digital x-ray images of 40 cadavers during dissection classes: dissection was combined with visualisation and palpation of landmarks, organs and structures on themselves, colleagues and the cadaver. The use of such images helps students to develop expertise to interpret radiological images, magnetic resonance (MR) or computerized tomography (CT) by reconstructing a two-dimensional image into a three-dimensional image. Furthermore, they need to learn to correlate these images with a living person in clinical practice. A virtual reality jigsaw for learning anatomy has also been reported to assist with understanding complex 3D structures and engaging students [53]. The disadvantage of 3D technology is that interacting with 3D models with a mouse does not convey a sense of touch or weight [23].

While some believe that anatomy should be taught using multimedia, problem-based learning and computer imaging [54], others challenge this believing that these media should supplement rather than supersede dissection [15, 55]. Technological teaching aids are not currently being used very much in Australia or New Zealand [13].

6. Learning Outcomes

Of central importance is student learning of Anatomy. Student engagement requires both an interested, passionate (stimulated) teaching style and the provision of meaningfully learning activities involving student peer interactions. A recent study has compared student engagement levels between 2 different styles and strategies for teaching first year Anatomy: health science students taught with a stimulated (passionate) teaching style with active, guided, self-directed experiential learning strategies using these new approaches versus medical students taught with more traditional didactic teaching style and strategies [18, 19]. Health science students using the new approaches rated engagement variables significantly higher ($p<0.05$) than their medical counterparts using more didactic approaches: their overall enjoyment of Anatomy; overall quality of Anatomy learning experiences; general level of interest in Anatomy teaching activities; importance of Anatomy learning activities to later years of their course and future professional career; and overall level of interaction with both peers and teachers. Health science students also had a significantly stronger preference for small group learning, less structured learning and more self-directed learning compared to medical students. Furthermore, while health science students demonstrated a greater than 80% increase in pass rates and a greater than 30% decrease in fail rate since 2005, grades for medical students remained unchanged during this time [18, 19]. In addition, 83% of the health science students thought the practical, innovative approaches assisted their deeper understanding of anatomy, 81% thought they assisted their long-term memory of anatomy, and 82% reported that the learning and memory skills they acquired in anatomy have been useful in other subjects.

This study suggested that passion and enthusiasm of teaching staff is a prime factor in enhancing student engagement and interest in anatomy. Using these new teaching approaches, students are not only fully engaged, but also able to achieve better results due to undertaking meaningful and worthwhile tasks. A clear improvement in grades suggests that these new teaching approaches have led to an improvement in anatomy knowledge. Specifically, it appears that whiteboarding, play doh, movement and surface anatomy (principally body painting) provide visual and tactile cues that appear to help with long term retention of information and deep learning. These innovative approaches appear to be producing highly motivated and self-directed learners with improved cognitive learning skills, commitment and a high level of (long-term) anatomy knowledge. This report suggests that a purely didactic environment diminishes students' appreciation, enjoyment and overall learning experiences in anatomy. The integration of these teaching approaches with wet anatomy classes and a team of well-trained tutors appears to provide motivation that leads to deeper learning.

While supporting and assisting first year anatomy students to develop their knowledge and learning skills in anatomy, these new teaching and learning approaches also appear to produce *self-directed, life-long learners.* This is consistent with previously reported self-directed learning approaches [39, 51, 52, 56-63], however, this recent study offers the integration of a more widespread variety of teaching approaches than those reported previously.

Qualitative results obtained from students using these new teaching and learning approaches suggests that students enjoy the positive, enthusiastic and interactive teaching style. Students reported the classes to be entertaining, and they enjoyed the hands-on learning experiences of whiteboarding, using cadavers and body painting [18, 19]. Positive feedback is also a strategy that works well in developing the confidence and self-directed learning abilities of first year anatomy students. Through positive feedback students learn to take responsibility for their own learning and realistically assess their abilities and progress. When students take on personal commitment their self-confidence regarding their capabilities improves; this is significant as there is a high correlation between self-concept and achievement [64].

CONCLUSION

Taking these new perspectives on traditional methods of teaching anatomy provides a much more engaging, motivating, inspiring and enjoyable environment for student learning. New teaching and learning approaches for anatomy produce students that can achieve the desired learning outcomes, and are confident, self-direct learners. When used in conjunction with wet anatomy classes, these innovative approaches appear to provide motivation that leads to deeper learning. However, where cadaveric tissues are not available, these techniques can be used in isolation for self-directed study. Using these approaches, students appear to be enjoying their anatomy experience in their first year at University and they are displaying levels of engagement and achievement not seen before in this area. Many students are so engaged and inspired by their learning experience that they return to work as casual staff in Anatomy as tutors and dissectors. Such outcomes are encouraging and suggest a productive path to revolutionize the teaching of Anatomy in ways that meet students' needs and the challenges of the contemporary context.

REFERENCES

[1] Kift, S. Enhancing First Year and Easing Transition: A Learner-Centred Approach. *Teaching Matters University of Tasmania*. 2004.

[2] Shneiderman, B. Education by Engagement and Construction: Can Distance Education be Better than Face-to-Face. *URL: http://www. hitl. washington. edu/scivw/EVE/ distance. htm* 1994, [accessed January, 2012].

[3] Shneiderman, B; Alavi, M; Norman, K; Borkowski, E. Windows of opportunity in electronic classrooms, *Communications of the ACM,* 1995: 38(11), 19-24.

[4] Kearsley G. The Virtual Professor. A personal case study. URL: http://home. sprynet. com/~gkearsley/virtual. htm [accessed January, 2012]. 1997.

[5] Gilmore, H. Higher Education. Nearly 20% of university students don't last full year. *Sydney Morning Herald*, 2009: 6 December.

[6] Biggs, J. Constructing learning by aligning teaching: constructive alignment. *Teaching for quality learning at university: what the student does.* 2003: Second edition. Buckingham: SRHE and Open University Press.

[7] Collins, TG; Given, RL; Hulsebosch, CE; Miller, BT. Status of gross anatomy in the U. S. and Canada: dilemma for the 21st century. *Clin. Anat.*, 1994: 7, 275-296.

[8] Fasel JH, Bader C, Gailloud P. Anatomy teaching for medical undergraduates: general practice as a guideline? The brain. *Clin. Anat.*, 1999: 12, 115-119.

[9] Leong SK. Back to basics. *Clin. Anat.*, 1999: 12, 422-26.

[10] Creswell A. Student doctors missing Anatomy lessons. *The Australian*2010: March 31.

[11] Heylings DJA. Anatomy 1999-2000; the curriculum, who teaches it and how? *Med. Educ.*, 2002: 36, 702-10.

[12] Leung KK, Lu KS, Huang TS, Hsieh BS. Anatomy instruction in medical schools; connecting the past and the future. *Adv. Health Sci. Educ.*, 2006: 11, 209-215.

[13] Craig, SJ; Tait, N; Boers, D; McAndrew, DJ. Review of anatomy education in Australia and New Zealand medical schools. *ANZ Journal of Surgery*, 2010: 80, 212-216.

[14] Al-Shaqsi, S; Stringer MD. Do senior medical students know enough clinical anatomy? *NZMJ*, 2011: 124, 113-116.

[15] Ellis H. Teaching in the dissecting room. *Clin. Anat.*, 2001: 14, 149-51.

[16] McKeown, P; Heylings, D; Stevenson, M; et al. The impact of curricular change on medical students' knowledge of anatomy. *Med Educ*. 2003: 37, 954-61.

[17] James D, Purkayastha, S, Paraskevas, P, Shafiq, O, Darzi A, Athanasiou T. Anatomy: the future teaching of undergraduates. *Hosp. Med.*, 2004: 65(11), 681-5.

[18] Diaz, CM. Stimulating multidisciplinary first year students to learn Anatomy for life via innovative, pro-active approaches to improve engagement and learning outcomes. *ALTC Citations for Outstanding Contributions to Student Learning, Australia. 2011.*

[19] Diaz, CM, Woolley, T. Engaging multidisciplinary first year students to learn anatomy via stimulating teaching and active, experiential learning approaches. 2013: Submitted.

[20] Hattam, A; Diaz, CM. Medical student-initiated anatomy education: an extracurricular experience at a regional medical school. *MJA*, 2012: 197(4), 218.

[21] Hattam. Incorporating clinically relevant radiology teaching into a medical student-initiated anatomy education program. *J. Med. Imag. Rad. Oncol.,*2013: in press.

[22] Pederson, K. Supporting Students with Varied Spatial Reasoning Abilities in the Anatomy Classroom. 2012: *Teaching Innovation Projects*, 2(1), article 2.

[23] Brenton, H; Hernandez, J; Bello, F; Strutton, P; Purkayastha, S; Firth, T; Darzi, A. Using multimedia and Web 3D to enhance anatomy teaching. *Comp. Educ.,* 2007: 49, 32–53.

[24] Moxham BJ and Moxham SA. The relationships between attitudes, course aims and teaching methods for the teaching of gross anatomy in the medical curriculum. *Eur. J. Anat.*, 2007: 1, 19-30.

[25] Race, P. The lecturer's toolkit. A practical guide to assessment, learning and teaching. *Routledge, Taylor and Francis*. 2007.

[26] Kolb DA and Fry R. Toward and applied theory of experiential learning; in C. Cooper (ed)Theories of Group Process, London: John Wiley. 1975.

[27] Itin CM. Reasserting the philosophy of experiential education as a vehicle for change in the 21st century. *J. Exp. Educ., 1999:* 22(2), 91-98.

[28] Dunn, L; Morgan, C; O'EReilly, M; Parry, S. The student assessment handbook: nerw directions in traditional and online assessment. 2004: *London, Routledge*.

[29] Kearsley G. and Schneiderman B. Engagement Theory: A Framework for Technology-Based Teaching and Learning. *Educ. Technol.,* 1998: 38(5), 20-23.

[30] Kuh G. D. 2009. High impact activities and implications for curriculum design. *Queensland University July*.

[31] Kift S. 2009. Articulating a transition pedagogy to scaffold and to enhance the first year student learning experience in Australian higher education. *Final Report for ALTC Senior Fellowship Program.*

[32] Stephens CD 2000. Forget the sailboard--let's go whiteboarding! *Dent Update*. 2000: 27(5), 236-40.

[33] McMenamin, PG. Body Painting as a Tool in Clinical Anatomy Teaching. *Anat. Sci. Educ.,* 2008: 1, 139–144.

[34] Azer, SA. Learning surface anatomy: which learning approach is effective in an integrated PBL curriculum? *Med. Teach*, 2011:33(1), 78-80.

[35] Clavert, P; Bouchaïb, J; Duparc, F; Kahn, JL. A plea for the use of drawing in human anatomy teaching. *Surg. Radiol Anat.*, 2012: 34(8), 787-9. Epub 2012 Jun 16.

[36] Nanjundaiah K and Chowdapurkar S. Body painting: A tool which can be used to teach surface anatomy. *J. Clin. Diagn., 2012: Res*. 1-4.

[37] Collins, JP. Modern approaches to teaching and learning anatomy. *BMJ*, 2008: 337, 665-667.

[38] Patel KM and Moxham BJ. Attitudes of professional anatomists to curricular change. *Clin. Anat.*, 2006: 19, 132-41.

[39] Kang SH, Shin JS, Hwang YI. The use of specially designed tasks to enhance student interest in the cadaver dissection laboratory. *Anat. Sci. Educ.* 2012:5(2), 76-82.

[40] Hagens, G. Gunther Von Hagens Bodyworlds anatomical exhibition of real human bodies. URL: www. koerperwelten. de/en/pages/gunther_von_hagens. asp. 2004.

[41] Davies, M. Concept mapping, mind mapping and argument mapping: what are the differences and do they matter? University of Melbourne, *High. Educ*. 2010.

[42] Biggs, J. Student approaches to learning and studying. Hawthorn, Vic. *Australian Council for Educational Research (ACER*), 1987.

[43] Paivio, A. Imagery and verbal processes. New York: Holt, Rinehart, and Winston. 1971.

[44] Biktimirov, EN; Nilson, LB. Show Them the money: Using mind mapping in the introductory finance course. *J. Fin. Educ.,* 2006: 32, 72–86.

[45] Kleiner A, Keating J, Wise K, Haines T. The fine art of anatomy. Research at Monash, *Monash University*. 2004.

[46] Royal Australian College of Surgeons. Anatomedia program revolutionizing anatomy teaching. *Surg. News*, 2004: 5, 14.

[47] N. Eizenberg, C. Briggs, P. Barker and I. Grkovic. Anatomedia: a New Approach to Medical Education Developments in Anatomy. "ANZ J Surgery Review", *Austr. NZ J. Surg*., 2006: 76, 709.

[48] Jastrow H, Hollinderbaumer A. On the use and value of new media and how medical students assess their effectiveness in learning anatomy. *Anat. Rec*., 2004: 280, 20-29.

[49] Hallgren RC, Parkhurst PE, Monson CL, Crewe NM. An interactive, web-based tool for learning anatomic landmarks. *Acad. Med.,* 2002: 77, 263-265.

[50] Spitzer, VM; Scherzinger, AL. Virtual anatomy: an anatomist's play- ground. *Clin. Anat.* 2006:19,192–203.

[51] Wright SJ. Student perceptions of an upper-level, undergraduate human anatomy laboratory course with cadavers. *Anat. Sci. Educ.* 2012:5(3), 146-57.

[52] Kotzé SH, Mole CG, Greyling LM. The translucent cadaver: an evaluation of the use of full body digital x-ray images and drawings in surface anatomy education. *Anat. Sci. Educ.* 2012: 5(5). 287-94.

[53] Ruthebeck, GS, Carati CJ, Gibbins IL, Reynolds KJ. A virtual reality 3D jigsaw for teaching anatomy. *Stud. Health Technol. Inform*., 2008: 132, 436-8.

[54] McLachlan JC, Bligh J, Bradley P, Searle J. Teaching anatomy without cadavers. *Med. Educ*., 2004: 38, 418-24.

[55] Mitchell, B; Stephens, CR. Teaching anatomy as a multimedia experience. *Med. Educ.,* 2004: 38, 911-12.

[56] Evans DJR, Cuffe T. Near-peer teaching in anatomy: An approach for deeper learning. *Anat. Sci. Educ.,* 2009: 2, 227-233.

[57] Havet E, Duparc F, Peltier J, Tobenas-Dujardin AC, Fréger P. The article critique as a problem-based teaching method for medical students early in their training: a French example using anatomy. *Surg. Radiol. Anat.* 2012: 34(1), 81-4.

[58] Findlater GS, Kristmundsdottir F, Parson SH, Gillingwater TH. Development of a supported self-directed learning approach for anatomy education. *Anat. Sci. Educ.* 2012: 5(2), 114-21.

[59] Inuwa IM. Perceptions and attitudes of first-year medical students on a modified team-based learning (TBL) strategy in anatomy. *Sultan Qaboos Univ. Med. J.* 2012: 12(3), 336-43.

[60] Khalil MK, Nelson LD, Kibble JD. The use of self-learning modules to facilitate learning of basic science concepts in an integrated medical curriculum. *Anat. Sci. Educ.* 2010:3(5), 219-26.

[61] Minhas PS, Ghosh A, Swanzy L. The effects of passive and active learning on student preference and performance in an undergraduate basic science course. *Anat. Sci. Educ.* 2012: 5(4), 200-7.

[62] Kotur PF. Introduction of evidence-based medicine in undergraduate medical curriculum for development of professional competencies in medical students. *Curr. Opin. Anaesthesiol.,*2012: 25(6), 719-23.

[63] Van Schaik S, Plant J, O'Sullivan P. Promoting self-directed learning through portfolios in undergraduate medical education: The mentors' perspective. *Med. Teach.* 2012: Oct 26 [Epub ahead of print].

[64] Dweck, CS. Self theories: their role in motivation, personality and development. *Lillington, NC, Taylor and Francis, 2000.*

In: Education in Anatomical Sciences
Editor: Paul Ganguly

ISBN: 978-1-62618-488-6

Chapter 13

AN ANATOMY LEARNING RESOURCE CENTER: WHAT SHOULD IT LOOK LIKE AND WHAT SHOULD IT IDEALLY ACHIEVE?

Paul G. McMenamin* and Colin McHenry
Department of Anatomy and Developmental Biology
School of Biomedical Sciences
Faculty of Medicine, Nursing and Health Sciences
Monash University, Clayton
Victoria, Australia

DEFINITION

When the author refers to an anatomy or anatomical sciences learning resource centre ('Anatomical Sciences Learning Resource Centre') he is visualizing a real teaching space which may or may not be augmented by a virtual space and resources. It is pertinent when considering the scope of such a centre to establish whether one is referring to 'anatomy' in its broadest context – probably more correctly called 'anatomical sciences' (development/ embryology, histology, neuroanatomy, and topographical or 'gross' anatomy) or in the more restrictive sense of topographical anatomy alone. In this chapter I will primarily focus on human topographical anatomy but will give examples of how a resource centre could easily be expanded to include resources that cover the broader scope of the discipline.

The Need for an Anatomy Learning Resource Centre

Most anatomy 'departments' or 'schools', besides possibly having a dissecting facility, usually have some sort of teaching space that students use either for formal classes or more self-directed learning. This space may range from a simple area with a few plastic models and

* E-mail: paul.mcmenamin@monash.edu.

charts to a more complex space rich in learning resources ranging from museum-like permanent displays which may contain bottled specimens or plastinated specimens augmented by computer laboratories. Wherever you are on this scale I hope this chapter gives you food for thought about how to design a space if you do not have one, how to make the best of teaching resources if you do have one and perhaps for those who 'have it all' to provoke them into considering whether there is still room for enhancing what they already have established.

Who Are You Teaching and How It May Influence Design

The discipline area(s) students being taught in your department will greatly influence the nature of the material you will have to include in a resource centre and how it is presented. For example, you may teach medical students only, but I imagine that is becoming increasingly rare in modern universities. Most anatomy 'departments' or 'schools' also teach various allied health professional students who will have a range of needs in terms of depth of understanding and content of anatomy, from nursing and paramedics, to specialist groups such as physiotherapists, dentists, podiatrists and osteopaths/chiropractioners. Furthermore, besides health professionals there are many departments, including ourselves at Monash University, that teach into biomedical science or science courses. Teaching these science based curricula offers a lot of flexibility in terms of the 'context' in which material can be presented. This will be explored more fully later in the chapter.

Another factor to be taken into consideration when designing a resource area is whether your students are 'undergraduates' (ie entering university directly from high school) or 'postgraduates' (namely those who have a previous degree), which may or may not include some sciences, who are now undertaking a professional qualification such as medicine. You may even want to consider catering for qualified practitioners who wish to up skill or undergo refresher courses to enhance or enrich anatomy knowledge specific to their specialty.

What Should an Anatomy Learning Resource Centre Contain?

The answer to this is influenced by the response to the questions raised above. If for example a 'department' or unit is only commissioned to teach medical students then they would have a distinct set of parameters in mind and the displays and specimens could focus purely on detailed topographic anatomy of clinical relevance. If on the other hand a more general and less detailed curriculum is being offered to allied health professionals then the displays would have to reflect this.

On the same theme if, for example, physiotherapy students are an important cohort, then the quality of the musculoskeletal system specimens and displays would be greater than perhaps needed by a centre catering to nursing students. Equally, if students are studying anatomical sciences as part of a generalized science or biomedical science degree then the displays and themes may be much broader.

These issues must be bore in mind when considering which of the ideas below are most applicable to you.

Anatomical dissections are a self-evident need for any anatomy learning resource centre. There are several ways that professionally dissected (prosections) can be displayed besides the traditional bottled preparations (Figure 1) which have been used for over 200 years. Bottles (sometimes referred to as 'pots') may still have a role but they can give a teaching area a 'feel and look' of an 18th century curio museum exhibition (Figure 2). Bottles can also be heavy and difficult to move, especially larger specimens. In addition, the fluid needs to be maintained to ensure they do not become cloudy and discoloured.

Anatomical dissected specimens can be stored wet or refrigerated and placed on display when needed (Figure 3), although they cannot be left out too long or they dry out, an issue that must be borne in mind if you have large and multiple classes needing the same specimens. In addition, if one does display wet prosections then the area must have appropriate air handling capacity or air conditioning to avoid the build up of formaldehyde containing fumes. The *American Association of Anatomists* has excellent safety guidelines in its website in regard to handling of formaldehyde in laboratory settings (http://www.anatomy.org/content/gross-anatomy-laboratory-design).

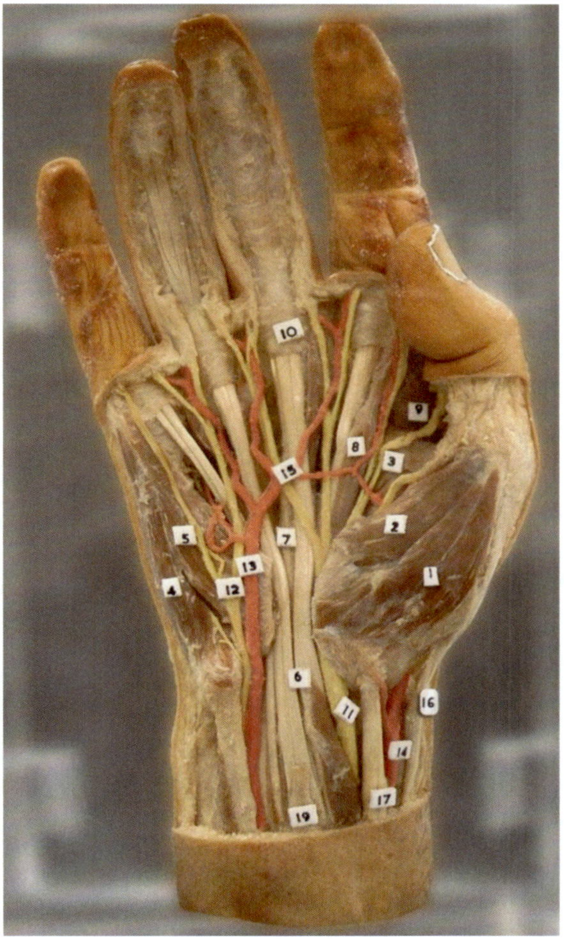

Figure 1. A bottled specimen of a prosected hand (vessels and nerves augmented in this photograph).

Plastinated specimens will enhance any anatomy teaching area. Plastination is the method of polymer embedding developed by Gunther von Hagens (1979, 1981), and made popular as part of '*Body Worlds*' and other similar exhibitions which have travelled the world and attracted millions of visitors – evidence in its own right of the fascination with human anatomy.

Figure 2. The 'Museum' space at Monash University, Melbourne, in 2010, before the redesign phase. Note the tables are simply covered with bottles and lack any obvious organization.

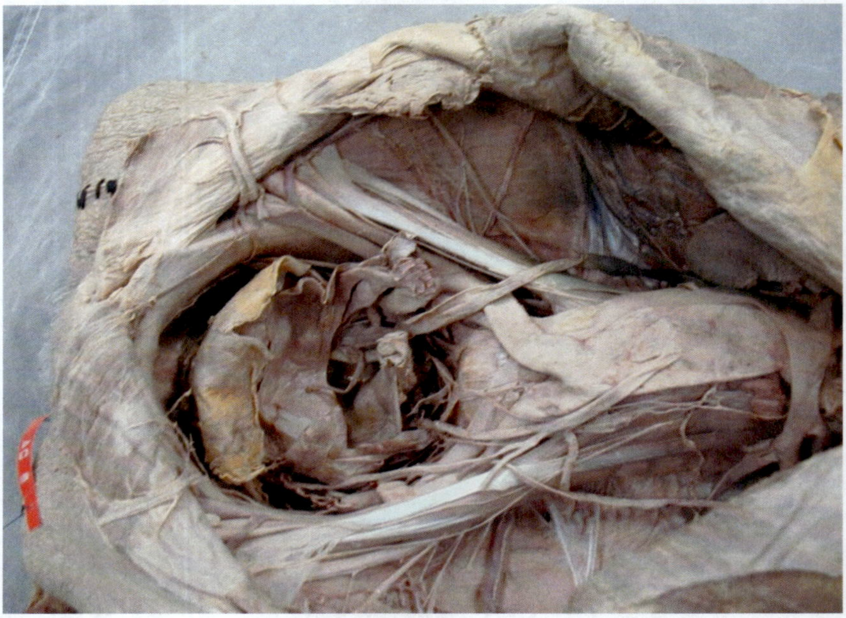

Figure 3. A traditional 'wet' prosection (posterior abdominal wall and pelvis).

There has been considerable debate about the ethics of displaying human bodies in unusual poses (is it education or entertainment) and about the ethics of using cadavers whose source has been the subject of some controversy (Collier, 2010). Plastination as a method of displaying human anatomical specimens has replaced bottled specimens in many departments. Plastinated specimens (Figure 4) are highly durable and have several advantages (Table 1) including ease of handling. However, the bureaucracy and extremely high cost of setting up a plastination suite, which arise partly due to safety issues, has in our experience been a major hurdle and forced us to look for alternatives (see below).

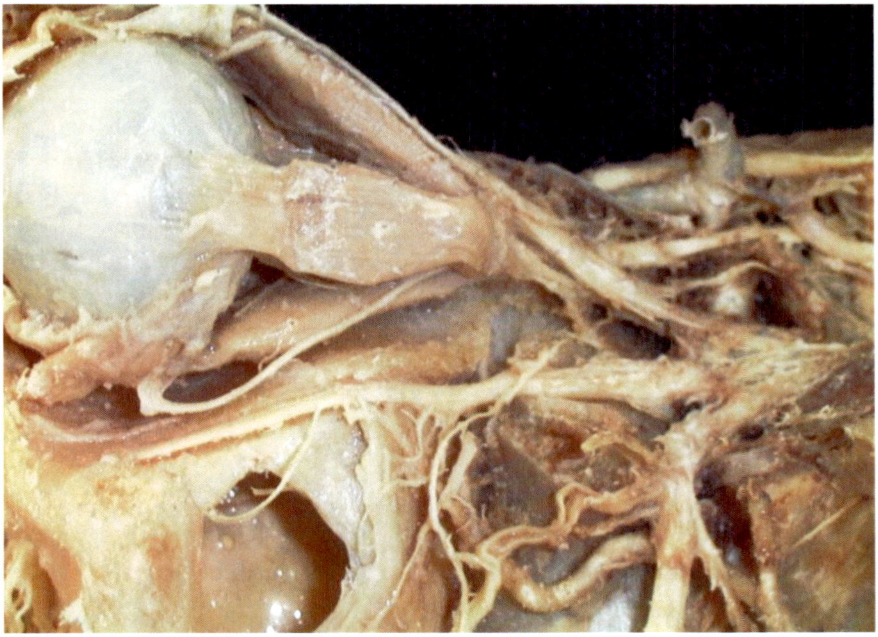

Figure 4. A plastinated specimen of the left orbit viewed from the lateral aspect.

Institutions may vary in their requirements for health and safety measures but most are likely to have some concerns about the large volumes of acetone, which although kept for most of the time at around -26°C still presents a theoretical explosive risk before getting to that temperature and during change overs.

Purchasing plastinated specimens from commercial suppliers also presents difficulties, partly due to cost, but also due to local 'Anatomy Acts'. In Australia, many state 'Anatomy Inspectorates' are not in favor of the importation of human remains from other countries. So the message is – before planning on setting up a plastination laboratory be cautious about the investment of time and money. However, if setting up costs can be kept to a reasonable level then this method is a very attractive way to produce specimens for a modern anatomy learning resource centre.

An ancient method of displaying anatomy: moulages. Moulages (wax models) were used for the teaching of anatomy from the late 1600s, originating in Florence. These moulages, often displayed in unusual artistic poses and partially 'clothed', were a much more pleasant way of learning anatomy than the alternative of dissecting unfixed cadavers. The technique spread to medical schools in Europe, Russia, North and South America, and Japan. Their creation is considered nowadays to have been a blend of early 'contemporary' art and very high craftsmanship (Cooke, 2010).

Charts and visual displays are an obvious asset to any teaching space and have two roles in the authors' view. Firstly, as an educational resource, but secondarily they can add to the aesthetic appeal of a teaching area. Perhaps in a more imaginative vein one could design the space so that it has changing visual displays or large interactive screens linked to popular anatomical images from your own collections.

Even static displays showing visually appealing images including histological material (which often has an intrinsic artist beauty), radiological material or cross sections can not only give a display area a modern artistic appeal, but if chosen wisely may also contain lots of interesting anatomy that can be discussed with students.

Computer aided learning: Popular software programs are an essential component of any modern teaching resource area. There are a number of excellent anatomy educational 'resources (Table 2) that offer varying degrees of interactivity and instructional text (such as 'An@tomedia'). Some are merely 'eBooks' which reproduce text and images that are already in the hard copy textbook. Others, such as 'VH Dissector' have lots of anatomical information but lack a student orientated guide or text with the images. They are best used by a teacher with a small or large class of students. The cost of these resources can vary dramatically from being free on the internet to being expensive for site licences.

Virtual Cadavers

One recent development that would greatly augment any learning resource centre is 'Anatomage Table' or virtual dissection table (http://www.anatomage.com). The designers of the table describe its value in the following manner:

> "Anatomage offers a unique, life-size interactive anatomy visualization table for the medical community. The Table offers an unprecedented realistic visualization of 3D anatomy. Delivering accurate anatomic details, it can be very complementary for cadaver based dissection courses. The Anatomage Table allows students to visualize skeletal tissues, muscles, organs and soft tissue that can be customized by virtually slicing, layering, and segmenting the anatomy. This adds a new dimension of depth to the education students receive. The Anatomage Table allows for a digital and septic interactive learning experience.

Having not seen the table myself, it is difficult to judge its short term and long term value. Since it is based on a similar principle to the 'VH Dissector' software from the 'Virtual Human Project', I am sure it will prove highly popular and if institutions have the resources, it would be a valuable asset to student learning areas. How it will look in 10 years time when technology will have advanced is unknown.

Table 1.

Method of Display	Advantages	Disadvantages	Cost effectiveness
Bottled specimens	Easy to prepare, prepare once, Longevity Can be labeled in bottle	Heavy Potential fire hazard Inconvenience to move Students cannot touch or handle specimen	Low cost, last for decades
Wet prosections	Easy to prepare, Can be stored until needed Can be handled by students, Structures mobile	No labels Storage – refridgeration preferred, Easily damaged with repeated use means may need to be replaced every 4-6 years Multiple copies may be needed to teach large classes	Higher due to storage costs and need to replace specimens
Plastination	Resilient and durable Can be physically handled by students	Specialist plastination suite, including freezers, vacuum chambers are required Health and safety issues with institution authorities due to large volumes of acetone required	Expensive to set up
Plastic models	Easy to purchase Easy to house/display Easy to move around	Not very anatomically accurate Not very detailed for many users	Can be expensive if multiple copies are needed
3D Printing	Multiple copies of prosections can be made easily	Although printed in mostly firm resin cured polymers printing can be made in upto 4 materials	Cost low once specimen is dissected and CT scanned
Computer based programs	Easy to buy 'off the shelf' Multiple levels of sophistication Appeal to younger generation	No interaction with a real specimen Experience is purely visual	Varies
Charts	Easy to obtain Cover multiple areas	2D diagrammatic Not real specimens Experience is purely visual	Low cost

Table 2.

Software	Available	Type	Cost (US $)
an@tomedia	CD/online	Instructional cadaver images Regional and systematic approach (not all modules are complete)	~$70-130 copy per disk (5 available)
VH Dissector	CD	Virtual human, interactive, great aid to axial (transverse) sectional anatomy	~$130 student copy ~$450 per instititionallicence
Primal Pictures	CD/DVD or online	Diagramatic 3D images, not real dissections	~$1000 per copy
A.D.A.M.	Online resource	Interactive anatomy, diagrammatic images, simplified anatomy suitable for allied health professionals, lacks details for medical teaching	~$100 for 12 month student subscription
Real Anatomy Software DVD (Wiley)	DVD	Real anatomy dissections, photographic atlas with ability to view from several perspectives	~$90 student copy

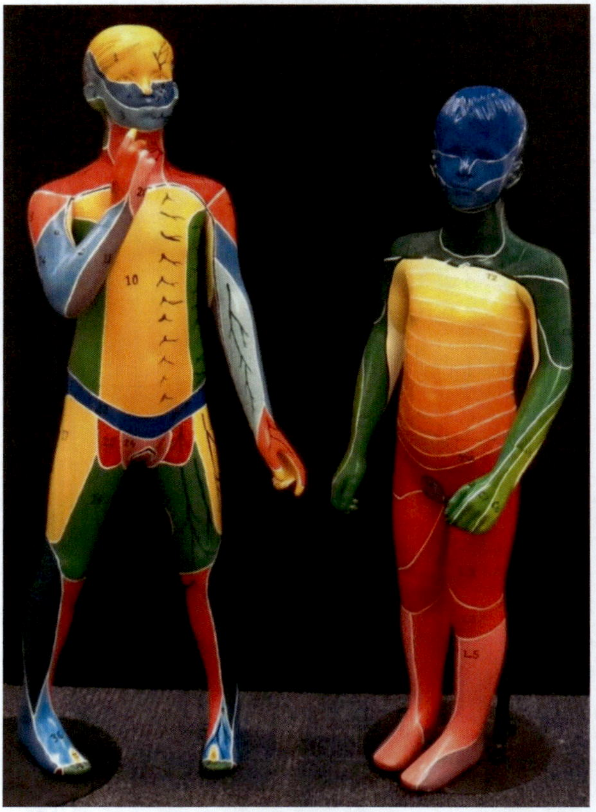

Figure 5. Mannequins painted to illustrate cutaneous nerve territories (left) and dermatomes (right).

Other Ideas for Displays

One concept that I have found many students struggle with is the difference between dermatomes and cutaneous nerve distributions. We get around this by illustrating these in three dimensions using mannequins. One can purchase cheap mannequins from commercial suppliers to clothing shops (from less than $100 to $200). Try to choose mannequins that are as anatomically accurate as possible. We selected the child mannequins as the size is more manageable in our resource centre.

After application of paint primer the next stage is to mark out one mannequin with the approximate dermatone map and then colour each dermatome with good acrylic paint and indicate with a label which dermatome it represents (Figure 5 right) and on the other map out the territories of the cutaneous nerve territories and the position of the actual cutaneous nerves (Figure 5 left).

We mount our mannequins on a stand with wheels that can be moved around the laboratory when required. The marking out and painting of the mannequins are good examples of projects where art students and medical/allied health students can collaborate on a defined project with a desired and valuable outcome.

Radiographic Resources

Clearly, in the modern era of medical imaging, it is critical to any good teaching resource centre that there is an appropriate range of radiological teaching material that accords with the role of the centre. In our case, we are responsible for teaching of normal anatomy. The domain of pathology is another department's responsibility. As such therefore we place an emphasis on normal radiological material from a range of imaging modalities including plain X-ray, CT, MR, vascular contrast, ultrasound and even tractography. Whilst hard copies, such as plain film X-rays, still have a minor role these can take up a large amount of space (namely they may require multiple light boxes), are time consuming to change for each class and take up considerable storage space. In our resource centre these teaching materials form the basis for weekly tutor-guided classes where the images are discussed. The images on are available (as 'Powerpoint' presentations) on the many computers in the lab for self-directed learning. This also means 3D animations files can be viewed by students, something that clearly is not possible on a light box. The choice of digital display versus hard copy display will depend on the available budget, but digital storage and archiving is much easier, and can be easily updated by collaborating with colleagues in the clinical settings.

A New Direction: 3D Printed Anatomical Specimens

3D printers produce physical color models quickly, easily and inexpensively from computer-aided design (CAD) and other digital data. Many companies have adopted 3D printing as a critical part of the iterative design process to aid the speed of creating prototypes from the concept stage. In our department we have adapted this technology to produce novel anatomical teaching specimens – 3D prints of copies of high quality prosections (Figure 6). This exciting development requires high quality prosections, which are then CT scanned at high resolution – a modern 64 slice CT scanner can produce pixel resolutions of 0.2 mm, which is sufficient for most prosections. The air tissue interfaces provide good contrast, allowing one to visualize even small nerves and vessels (Figure 7A).

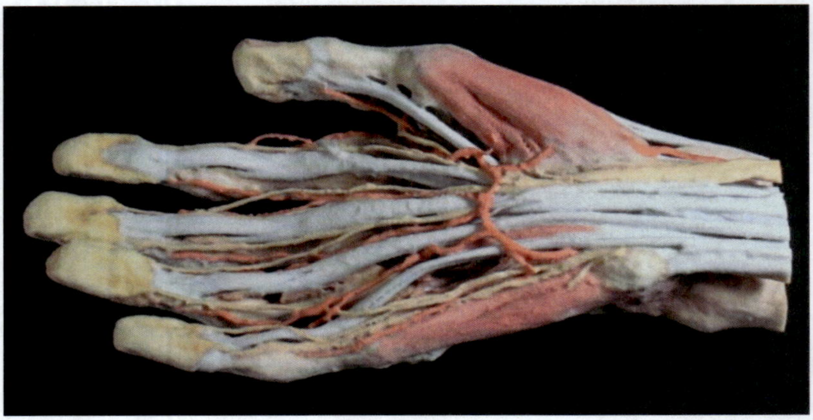

Figure 6. 3D print of a prosected hand.

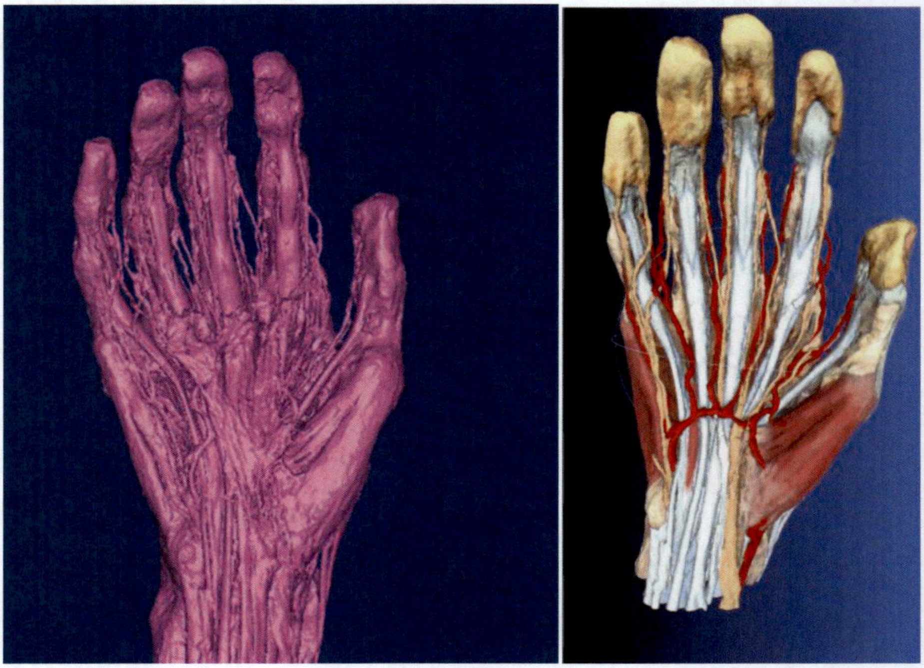

(a) (b)

Figure 7. (A) 3D visualization of CT-scanned prosection of hand and forearm. The 3D computer image is constructed from the CT data (in this case, exported from the scanner workstation in DICOM format) using image processing software (e.g. AMIRA, AVIZO, MIMICS, Simpleware, 3D Slicer), which creates a stereolithography file (.STL). (B) CT does not give information on colour, but anatomically realistic colour can be added using a package such as 3D Coat. The coloured STL file can then be printed in full colour as a 3D copy of the original prosection.

Once processed, a 3D stereolithography (STL) file can be then printed in a plastic/resin or plaster/gypsum material or indeed in multiple materials in the one specimen. The 3D object can be printed in plain white and colouring the model can be used as teaching activities or self-directed exercises. Alternatively, they can be printed in colour, which requires that the 3D STL file be digitally 'painted' using a software package such as '*3D coat*' (Pilgway) (Figure 7B).

There are a great many models of printers on the market with varying qualities of printing or resolution. We have chosen an additive printing method with resolution of each layer being in the 0.1mm range.

The three basic steps of 3D printing (use of Z650, 3DDD System given as an example):

1. A 3D CAD (computer aided design) file is imported into ZPrint™ software. The software slices the file into thin cross sectional slices, which are fed to the 3D printer.
2. The printer creates the model one layer at a time by spreading a layer of powder and inkjet printing a binder in the cross-section of the part.
3. The process is repeated until every layer is printed and the part is ready to be removed

The limitations of this technique are less of a concern than one may imagine, provided you have the right budget. For example, up to 4 different materials of different properties can be used to print. One need only browse the internet to see protype bicycles, and even cars, that have been 3D printed to begin to appreciate the possibilities for our discipline. We are only limited only by our imagination.

Self Directed Learning in an 'Anatomical Sciences Learning Resource Centre'

In the world of medical and allied health professional education with an ever increasing science and clinical knowledge base there is often less room in the curriculum to fit in what some perceive as the basic essentials as formal didactic teaching sessions in many of the pre-clinical sciences. Anatomy is no exception in this regard (a topic I am sure that has been dealt with elsewhere in this book).

Fortunately, in todays digital era we as teachers are no longer the sole custodians of knowledge and in many courses it behooves the student to take it upon themselves to cover some of the core curriculum. Thus self-directed learning is now, and was of course probably always, an essential part to any student's learning experience. One critical role of any 'Anatomical Sciences Learning Resource Centre' must be therefore to provide easy access to learning materials for students to access out of formal class contact time. The more engaging, exciting, and captivating the materials in the learning resource centre appear, then the more heavily utilized it will obviously be.

Issues such as ease of access, welcoming lighting, interesting visual displays, comfortable seating areas, wireless internet access and/or sufficient computer terminals will all increase the likelihood of students using the space in a worthwhile manner to further their studies.

In the author's experience the importance of senior members of academic staff wandering through this area and chatting to students about learning issues or difficulties can pay large dividends in terms of levels of student satisfaction with their course.

An obvious feature of these areas should be plenty of whiteboard space and group seating areas to facilitate group learning, especially if students are encouraged to form study groups. In one area of our anatomy teaching space we have recently used 'Idea Paint' (http://www.ideapaint.com/) to coat the entire wall surfaces. This is a quick and highly effective way of converting an otherwise useless wall into a large free drawing surface similar to a whiteboard.

Do We Need a Real Learning Resource Area – Why Not Just Have a Virtual Space?

There is the option, if an anatomy department or school does not have sufficient space to house a physical centre, that it could house a 'virtual' space where students would still find the learning materials they need for their course.

This has the advantage that it could be available 24/7 on the internet. There are very few obstacles to creating this form of learning resource and indeed there are many such resources already available on the internet although some require institutional logins. The use of *'You Tube'* and other free internet sites have only begun to be utilized, but are showing lots of promise (Jaffar, 2012). The disadvantages of a virtual space is that students will not interact with staff, nor will they learn in study groups or get the social peer-peer interactions that can enrich such spaces.

An example should suffice to illustrate this point. In our 'Anatomical Sciences Learning Resource Centre' we routinely have fine art students using anatomical specimens or osteological specimens as subjects for their drawings or paintings classes and it is not unusual to see art, science and medical students asking each other questions about either the anatomy being studied or their artworks/dissections. This richness of cross faculty fertilization is an ideal example of what universities should be about – broadening horizons and encountering experiences beyond the strict limits of your discipline.

Choosing an Overarching Theme for a Learning Resource Centre

There are a number of possible options when considering the manner in which a learning resource centre can be organized. One simple way would be to make it regional (upper limb, lower limb, thorax, abdomen, pelvis, and head and neck). The other obvious theme could be body systems (cardiovascular, gastrointestinal and so on). However, there are other options depending on the target audience and the breadth of what one considers 'anatomical sciences' as eluded to earlier. For example, if you teach significant numbers of biomedical and science students and have responsibility for histology, embryology/developmental biology and neuroanatomy then there are other exciting themes that could be picked up. One could devote an area to displays such as 'Fertilisation to Birth', 'Human Movement', 'Fuelling our body', 'The Special Senses' or other large concepts. Within these areas there could still of course be the learning materials required by special groups such as medical students.

There is of course one overarching concept that could be the basis for organization of any learning area in anatomical sciences and that is indeed the overarching law of biology – EVOLUTION. It would be a gross disservice to any university students or indeed a dereliction of duty as teaching academics if students studying anatomical sciences were not taught the big question - 'what is the evolutionary basis for human form and function?'.

This question is central to understanding the human body and nothing can be fully understood unless we appreciate the evolutionary history. At the entrance to our learning centre we have the 'Hillis plot' (Pennisi, 2003) prominently displayed (Figure 8) to provoke students to ask the 'Why?' questions. Whilst it is easy to, for example, teach the anatomy of limbs or vertebral columns as detailed in most textbooks, it is far more challenging to get your students to consider the functional adaptations to bipedalism in the human form and its evolutionary ancestry.

At Monash University are about to embark on a total redesign of our learning resource area (Figure 9) and it is the evolutionary theme, that whilst it may present the most challenging to design and implement, will we believe ultimately encourage students to ask the bigger questions as well as provide the platform for learning resources that will enable students to become captivated by the scientific underpinnings of anatomy which will greatly aid in the learning when compared to the more traditional view of rote memorizing lists.

Figure 8A. The re-designed new entrance to the Human Anatomy learning Resource Centre, Monash University, Melbourne, 2011. Note the circular Hillis plot on the wall at the entrance.

Anatomy Research

Lastly, and surely not least, it is important that our discipline tries to bring research being performed in our departments to the foreground for our students to appreciate that anatomy is not a 'dead' discipline, as indeed many, including academics in other biomedical sciences, think. This can be achieved by the regular display of posters from research groups or from student projects and perhaps housing seminar series in the facility.

Figure 8B. Refurbished anatomical Learning Resource Centre at Monash University, Melbourne, 2011 (the same space as shown in Figure 2).

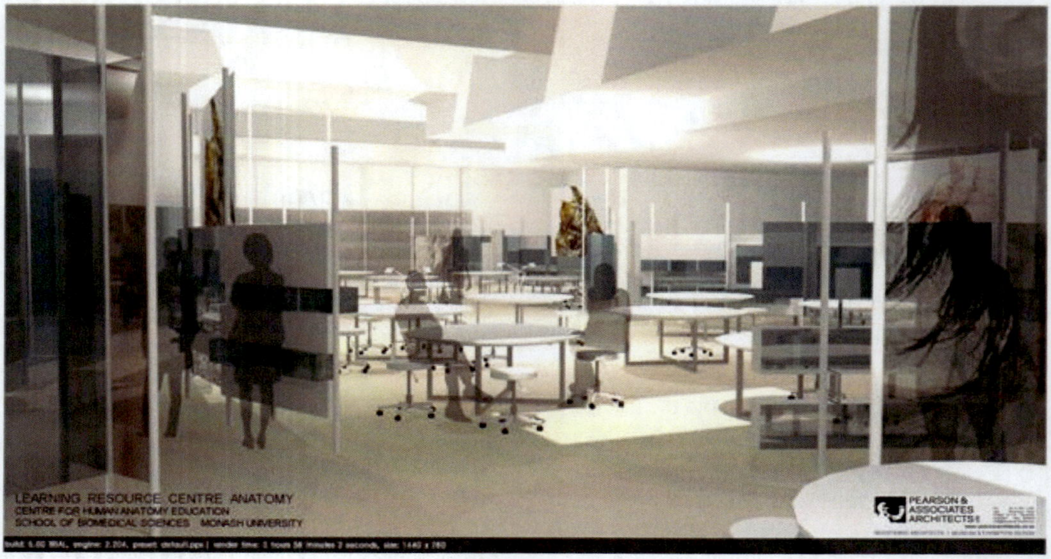

Figure 9. Part of the architect plans for the ASLC at Monash University.

Who Should Be Admitted to View Material in an Anatomy Learning Resource Centre?

This is a controversial issue with some local legislation governing who may be allowed to view human cadaveric remains. If the laws are not clear then it will remain the responsibility of the institution to determine who they think should be allowed to utilize the facility. In an era where over 10 million people have visited '*Bodyworlds*' or similar exhibits of plastinated human specimens and where there are internet sites with anatomy teaching materials it becomes, in the author's opinion, harder to justify restrictions in access. Many facilities only allow access to students studying anatomy.

(e.g. http://sydney.edu.au/medicine/anatomy/museums/wilson.php).

At Monash University we have open days in which the public is allowed access to the facility but they are not allowed to view wet prosections or the dissection room. There are bottled specimens and plastinated specimens and there is generally a great deal of interest and fascination with the material on display. There is always a staff member present when any groups come through the facility for a visit and we emphasize that people should be respectful to the situation.

REFERENCES

An improved method for the preservation of teaching specimens. *Arch. Pathol. Lab. Med.* 105(12):674-6.

Bickley HC, von Hagens G, Townsend FM. (1981)

Collier R (2010) Cadaver shows stir controversy. *CMAJ.;* 182:E687-8.

Cooke RA (2010) A moulage museum is not just a museum : Wax models as teaching instruments. *Virchows Arch.* 457(5):513-20.

Jaffar AA (2012) YouTube: An emerging tool in anatomy education. *Anat. Sci. Educ.* 5:158-64.

Pennisi E (2003) Modernizing the Tree of Life. *Science Vol* 300 no 5626; 1692-1697.

von Hagens G. (1979) Impregnation of soft biological specimens with thermosetting resins and elastomers. *Anat. Rec.* 1979 194:247-55.

In: Education in Anatomical Sciences
Editor: Paul Ganguly
ISBN: 978-1-62618-488-6

Chapter 14

WHAT IS THE BEST ASSESSMENT SYSTEM IN ANATOMY?

Sabri Kemahli*[*], *Aftab Ahmed Shaikh, Tasneem Fatima, Muhammad Atif Mazhar, Sadia Qazi and Paul Ganguly
College of Medicine, Alfaisal University, Riyadh,
Kingdom of Saudi Arabia

ABSTRACT

Medical education has undergone a lot of changes, especially in the last 30 years. Introduction of new education methods, tools and innovative curricula has necessitated a change in the assessment and evaluation, as well. As the move in curricula and educational methods has been towards integration and around students' activities, a change in the assessment and evaluation methods compatible with these was inevitable. Formative assessment, which is mainly intended to provide feedback and information about the progress, strengths and weaknesses of the students, should be an integral part of the assessment system. Summative evaluations should include continuous assessment, thus it should not be restricted to final examinations. All the assessment and evaluation methods and systems are required to be planned taking into account the psychometric properties and principles, namely, validity, reliability, fairness, practicality and efficiency. In order to assess various aspects of anatomy different methods are necessary. Planning the assessment should start with a well-structured exam blueprint. Since anatomy education in many medical schools is part of integrated curricula and learned/taught in a clinical context, the assessment and evaluation methods should comply with this. Thus methods such as objective structured clinical examination (OSCE)/objective structured practical examination (OSPE), in addition to MCQs with clinical vignettes are to be used. OSPE/OSCE stations should emphasize the clinical relevance of anatomical structures.

[*] e-mail address: skemahli@alfaisal.edu.

A. Assessment and Evaluation- Differences and Importance

Assessment and evaluation are used interchangeably in most cases, but there is a minor difference between them. Assessment requires the gathering of evidence of student performance over a period of time to measure learning and understanding. Evidence of learning could take the form of interviews, observations, written work, portfolios and examination along with many other learning tasks. Evaluation on the other hand, occurs when a mark is assigned after the completion of a task, test, quiz, lesson or learning activity. A mark in a course or clinical clerkship will determine if the student can perform the skills and has the knowledge for that area and would be seen as an evaluation. Assessment would be a review of journal entries, written work, presentation, research papers, essays, story writing, tests, exams etc. and will demonstrate a sense of more permanent learning and clearer picture of a student's ability. Although a student may receive high marks in a test, if he/she can't apply these in every day work, the high test marks (evaluations) matter little. "Assessment" is the process that a teacher/examiner conducts to measure how much a student/candidate has learned. "Evaluation" is the process used to measure how effective a particular approach or method is to achieve specified ends. So, although both aim the same thing, once a "value" (a numerical value or grade) is put, it can be called "evaluation".

The Purpose of Learning Anatomy as Medical Student and as a Medical Graduate

What is the outcome of anatomy education? What should be expected from a student who has completed his/her anatomy education? Anatomy is not just about listing or naming some structures, fossae, muscles, tendons, nerves, arteries, etc. It is much more important to correlate gross and microscopic structures with their functions and clinical relevance. So, anatomy is no longer rote learning or memorizing the anatomical structures. The students should learn them with their clinical relevance. Thus correlating anatomy with physiology *and* clinical sciences (including radiology) should be the aim. This aim has actually been realized by the implementation of integrated curricula, either with or without problem-based learning (PBL). At present students are exposed to learn anatomy in a clinical context in most medical schools. Thus it should not be a big challenge to organize assessment in an integrated way.

Besides, with the introduction of integrated curricula anatomy is no longer a discipline learned/taught in one year only. Anatomy education starts in year 1 but does /should continue till the final year of medical school. Therefore a continuous assessment/evaluation of anatomy should take place throughout the curriculum. In the first year(s) of medical school there is always a weight of anatomy in the curriculum. It is either in the form of traditional discipline-based ''anatomy'' course or integrated with all organ-system based courses or those such as ''structure and function'' courses. In the latter two, anatomy is learned and taught together with other disciplines such as physiology, embryology, biochemistry and sometimes in a clinical context, especially in problem-based curricula. As the assessment should be based on the intended learning outcomes, assessment should not only be based on factual knowledge but should contain or relate with clinical topics.

Problem-based learning requires students to be able to integrate basic and clinical sciences, scientifically based decision-making, clinical reasoning and psychomotor skills (Chakravarty et al., 2005). Assessment of anatomical understanding in PBL requires several types of examination that no single method of assessment can effectively test the three individual aspects of learning: knowledge, attitudes and skills (Nendaz and Tekian, 1999). "Learning objectives" in anatomy relate to both the structure and function of the body and are achieved through the health care problems (Chakravarty et al., 2005).

B. Types of Assessment

Assessment is the very important and integral part of the curriculum of any medical institute at undergraduate level and most of them are following the routine formative and summative type of assessment system. Michael Scriven was first to distinguish between the formative and summative assessment in the 1960's (Scriven, 1967). Formative assessment is a method for judging the progress and achievements and providing feedback while the program activities are in process. Summative assessment, on the other hand, is carried out at the end of the course or education and aims giving a final decision (grade, score, pass/fail decision) to the students. In other words, while former focuses on process the latter focuses on the outcome.

Assessment Drives Learning- What Is to Be Assessed Defines What Is to Be Learned

It is a well-known fact that students study according to the content of the examination. If a topic or concept is not evaluated it will not be learned by the students. Thus assessment/evaluation should be planned together with the planning of the course content.

C. What Should WE Assess? What Level in Miller's Pyramid?

Miller's pyramid traditionally only has 4 stages but during recent discussions there have been suggestions that we should include 2 underpinning awareness levels that usually occur before we "know". It is also important to consider them as these two new lower levels plays an important role in PBL curriculum. Each level of this pyramid can be measured by various methods. Needless to say, the methods differ according to the level, as well as the area to be assessed. The first two stages, 'knows' and 'knows how', are considered to correspond to the cognitive or knowledge area and can be assessed using the traditional assessment tools of written and oral tests. The tests can be given on paper or delivered by electronic means (internet or computer-based tests).

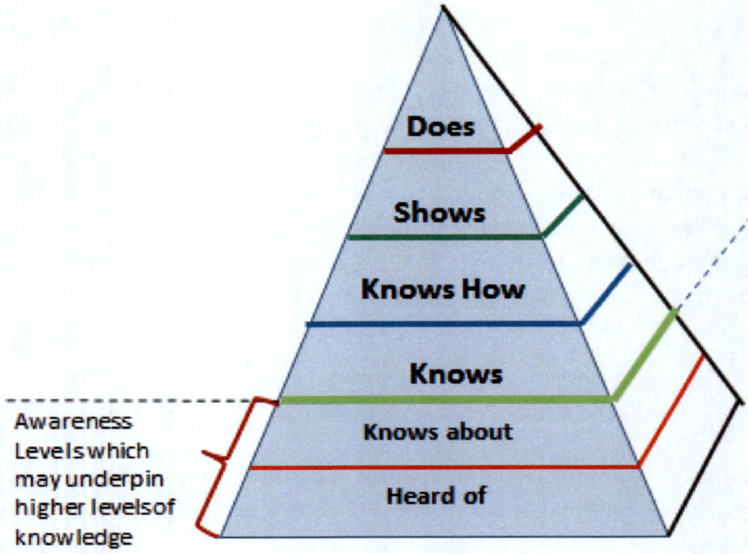

Figure 1. Miller's pyramid with two underpinning awareness levels.

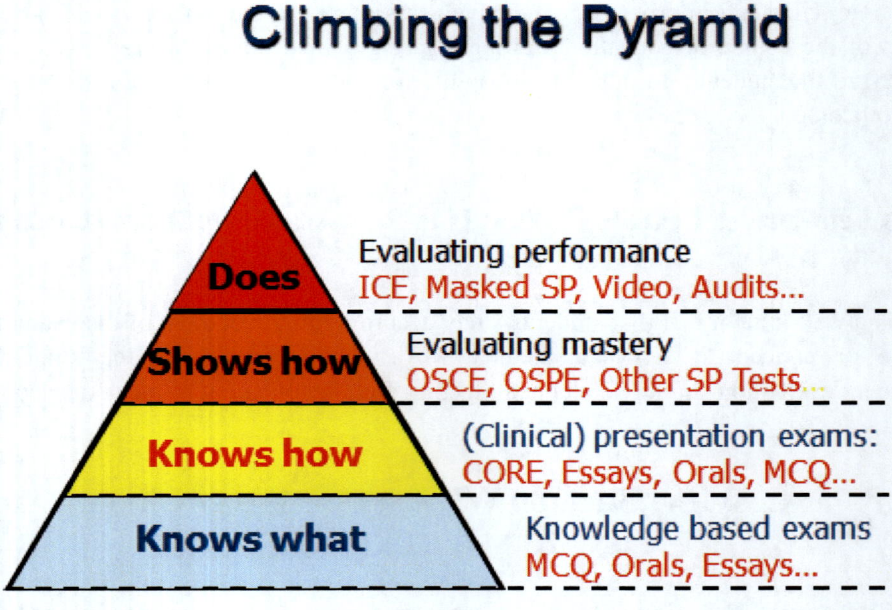

Figure 2. Miler's pyramid with related assessment/evaluation methods.

However, 'knowing' and 'knowing how' clearly do not necessarily extrapolate to the application of knowledge in the naming the structures, defining the paths, origins, listing and defining the related gross and microscopic structures. Such knowledge can easily be tested by written or computer-based tests. Although oral tests can also be used, it has the disadvantage of being very subjective and usually unstructured. If however, structured oral tests are developed they can be used to measure these areas.

The higher order of assessment which should measure how a student "shows" and "does" particular behaviour, skills and attitudes necessitates methods such as OSCEs and OSPEs. Objective structured practical examinations (OSPE) are already in use for a long time. There are schools who use OSCE (objective structured clinical examinations) for anatomy. Since anatomy is taught in a clinical context, some OSCE stations can be designed to include skills that necessitate the use of anatomy knowledge, especially in the later years of medical education.

D. Psychometric Properties of Assessment

1. Validity

Validity is defined as "the extent to which a test measures what it is supposed to measure", but it is more than that. Validity is concerned with the soundness, trustworthiness, or legitimacy of the inferences that were made on the basis of the obtained scores. In other words, it is the interpretation made from the test results. Is the information that I have gathered the right kind of evidence for the decision I need to make or the intended use? How sound is the interpretation of the information? How do we determine the validity of the assessment method or the test that we use?

Professional judgment is keystone to validity. This judgment is made by the user of the information (i.e. the instructor for classroom assessment).Traditionally, validity comes from three evidences: content-related, criterion-related and construct-related. There are different types of validity, namely content validity, construct validity, predictive validity and face validity (Amin 2003).

a) Content Validity

Content validity refers to the principle that the content of the examination should be representative of the area to be assessed. Thus the questions, OSCE/OSPE stations and other measurement tools should cover and represent as much part of the course content as possible. Since, everything cannot be included in the examination the more important points representing a wide area of the course should be included. Suppose we wanted to test for everything first year students learn in a four week module on reproduction. Can you imagine how long the test would be and how much time the students would take to complete the test? What you do is to select a sample of what has been taught, and use this student achievement as basis for judging that the students demonstrate knowledge about the module. Adequate sampling can be done by reviewing the match between the intended learning outcomes and what is on the test. This process begins with clear learning objectives and preparing a table of specification for these targets (Biggs 2001).

The table of specification or the test blueprint is a two-way grid that shows the content and types of learning targets. Table 1 is designed as a sample of blueprint for an integrated course with various disciplines, such as anatomy, histology, radiology and different methods of examination.

In this example main content areas to be assessed are;

1. Sexual differentiation and development of the female and male reproductive systems.
2. External genitalia
3. Puberty
4. The breast
5. Pelvis and Perineum
6. Structure, function of the female reproductive system and its regulation
7. Structure, function of the male reproductive system and its regulation

Table 1. An example of a blueprint of a test in Reproduction module

Learning objectives	MCQ 45%	SAQ 20%	EMQ 10%	OSPE 25%
<u>Male Reproductive System</u> Describe the scrotum, testis and epididymis including blood supply and innervation.	✓			✓
Discuss hormonal regulation of testicular function and the physiological effects of testosterone on male reproductive anatomy.	✓		✓	
Explain the structure of prostate gland, seminal vesicles and their anatomical relations.	✓			
Justify the position of prostate and its relationship to digital rectal examination for evaluation of various prostatic conditions.	✓			✓
Explain the descent of the testes into the scrotum.	✓	✓		✓
Defend that the body temperature is harmful for spermatogenesis.		✓		
List the key functions of Sertoli and Leydig cells			✓	
Describe the development of the reproductive tract in the male & the female during fetal life, its genetic and hormonal control and abnormalities.	✓			
Describe the anatomy of urethra.	✓			
Discuss the differences of male and female in its various parts	✓	✓		
Identify the parts of the urethra through which a Foley's catheter would pass in male and female.				✓
<u>Female Reproductive System</u> Describe the anatomy of the female reproductive tract and their anatomical relationships including peritoneal coverings.	✓			✓
Recognize the location of urogenital diaphragm, superficial and deep perineal muscles.				
Interpret the spread of carcinoma in the reproductive organs including breast with the knowledge on histology, anatomical relations, blood supply and lymphatic drainage.	✓	✓		✓
Discuss the relationship between the identified microscopic structures of male and female reproductive system and the functional aspects of the identified tissues.	✓			✓
Apply the knowledge to a few common clinical situations (subfertility, labour) to explain how the anatomy physiology and biochemistry are altered in the given situations.	✓	✓		
Interpret two-dimensional images of the Reproductive System from radiographic techniques.				✓
Describe reasons for infertility in the male and female and methods for assisted reproduction.	✓			
Describe the processes involved in a normal labour and delivery and some common problems of labour in relation to pelvis size.	✓			✓

b) Predictive Validity

Predictive validity is relating an assessment to other valued measure (criterion) that either provides an estimate of current (concurrent criterion-related evidence) or future performance (predictive criterion-related evidence) in a related area. The principle is that when we have two or more measures of the same thing, and these measures provide similar results, then we have established criterion-related evidence. For example, if the assessment of a student's skills in using a microscope through observation coincides with the student's score on a quiz that tests steps in using microscope, then there is criterion-related evidence that the inference about the skill of this student is valid.

Similarly, if we are interested in the extent to which preparation by our students, as indicated by scores on a final exam in anatomy, predicts how well they will do next year, then you can examine the grades of previous students and determine informally if students who scored high on your final exam are getting high grades and students who scored low on your final exam are obtaining low grades. If a correlation is found, then an inference about predicting how your students will performed, based on their final exam is valid, particularly, predictive criterion-related validity (Biggs 2001, Zhao 2011).

Another example can be to look for if the scores in an anatomy exam of the cardiovascular system is predicting the students' performance in cardiovascular physical examination.

c) Construct Validity

A construct refers to an unobservable trait or characteristics that a person possesses, such as intelligence, reading comprehension, honesty, self-concept, attitude, reasoning, learning style and anxiety.

These are not measured directly; rather the characteristic is constructed to account for behaviour that can be observed. Three types of construct-related evidence are theoretical, logical and statistical. Theoretical explanation is to define the characteristic in such a way that its meaning is clear and not confused with any other constructs like "what is attitude" or '"how much students enjoy reading". Logical analyses on the other hand can be done by asking the students to comment on what they were thinking when they answered the questions, or compare the scores of groups who, as determined by other criteria, should respond differently.

Finally, statistical procedures can be used to correlate scores from measures of the construct from other measures of the same construct and measures of similar, but different construct. For example, self-concept of academic ability scores from one survey should be related to another measure of the same thing (convergent construct-related evidence) but less related to measures of self-concept of physical ability (divergent construct-related evidence) (http://dante.udallas.edu, http://www.uwm.edu).

d) Face Validity

Face validity is the property of a test which shows what it is going to measure. An exam has a face validity if "it looks like" it is going to measure what it is supposed to. For example, if the exam is supposed to measure the knowledge of the students for gross anatomy of lungs, it should look like that it really will assess that area (Amin 2003).

2. Reliability

Like validity, reliability term has been used for so many years to describe an essential characteristic of sound assessment. Reliability is concerned with the consistency, stability, and dependability of the results. In other words, a reliable result is one that shows similar performance at different times or under different conditions.

But even if the scores are reliable; they are not without some degree of error. In fact, all assessments have error; they are never perfect measure of the trait or skill. The concept of error in assessment is critical to understanding reliability. Conceptually, whenever we see something, we get an observed score or result. This observed score is a product of what the true or real ability or skill is plus some degree of error:

Observed score = True score + error

Reliability is directly related to error. It is not a matter of all or none, as if some results are reliable and others unreliable. Rather, for each assessment there is some degree of error. Thus we think in terms of low, moderate, or high reliability. It is important to remember that error can be positive or negative.

That is, the observed score can be higher or lower that the true score depending on the nature of the error. For example, if the student is sick, tired, in bad mood or distracted, the score may have negative error and underestimate the true score (Nygard et al., 2009, Cohen and Swerdlik 2004).

Methods of Establishing Reliability Evidences

The test-retest method of establishing reliability is frequently used; that is giving the same test twice to the same students at two different points in time. Other methods include parallel-forms method and alternate-forms reliability estimates. Parallel forms of a test exist when, for each form of the test, the means and the variances of observed test scores are equal. Alternate forms are simply different versions of a test that have been constructed so as to be parallel, in which the two forms of the tests are typically designed to be equivalent with respect to variables such as content and level of difficulty (http://www.uwm.edu, Nygaard et al. 2009, Armstrong and Conard 1995).

To enhance reliability, the following suggestions are to be considered:

- Sufficient number of items or tasks should be used.
- Independent observers should be utilized who provide similar score on the same performances.
- Items and tasks should be constructed that clearly differentiate students on what is being assessed.
- Objectivity of assessment protocol and scoring should be maintained.
- Assessment should be continuous until results are consistent.
- The influence of extraneous events or factors should be decreased or avoided.
- Frequent short assessments should be used rather than fewer long assessments.

3. Fairness

An assessment is labelled as a fair assessment if it gives same chance to every student to demonstrate achievement and yields scores that are comparably valid from one person or group to another. If some students have an advantage over others because of factors unrelated to what is being taught, then the assessment is not fair. Thus, neither the assessment task nor scoring is differentially affected by ethnic background, gender, race or other unrelated to what is being assessed. The following criteria represent potential influences that determine whether or not an assessment is fair (Nygaard et al., 2009, Cohen and Swerdlik 2004).

a) Student Knowledge of Learning Targets and Assessment

A fair assessment is one in which it is clear what will and will not be tested and your objective is not to fool or trick students or to outguess them on assessment. Rather, you need to be very clear and specific about the learning target – what is to be assessed and how it will be scored.

b) Opportunity to Learn

This means that students know what to learn and then are provided ample time and appropriate instruction. It is usually not sufficient to simply tell students what will be assessed and the test them. You must plan instruction that focuses specifically on helping students understand, providing students with feedback on their progress, and giving students the time they need to learn.

c) Avoiding Stereotypes

Stereotypes are judgments about how group of people will behave based on characteristics such as gender, race, socioeconomic status and physical appearance. Though it is impossible to avoid stereotypes completely because of our values, beliefs and preferences, we can control the influence of these prejudices.

Avoiding Bias in Assessment Task and Procedures

Bias is present if the assessment distorts performance because of the students' ethnicity, gender, race, religious background and so on. Bias appears in two forms: offensiveness and unfair penalization.

4. Practicality and Efficiency

High quality assessments are practical and efficient. Because time is a limited commodity for teachers, factors like familiarity with the method, time required, complexity of administration, ease of scoring, ease of interpretation and cost should be considered (Cohen and Swerdlik 2004).

a) Familiarity with the method

This includes knowing the strengths and limitations of the method, how to administer, how to score and interpret responses. Otherwise, teachers risk time and resources for questionable results.

b) Time required.

Gather only as much information as you need for the decision. The time required should include how long it takes to construct the assessment, and how long it takes to score the results. Thus, if you plan to use a test format (like multiple choice) over and over for different groups of students, it is efficient to put in considerable time preparing the assessment as long as you can use many of the same test items each year of the semester.

c) Complexity of administration.

The directions and procedures for administration should be clear and that little time and efforts are needed. Assessments that require long and complicated instructions are less efficient and because of probable students' misunderstanding, reliability and validity are affected.

d) Ease of scoring.

It is obvious that objective tests are easier to score than other methods. In general use the easiest method of scoring appropriate to the method and purpose of the assessment. Scoring performance-based assessment, essays and papers are more difficult to score so it is more practical to use rating scales and checklists rather than writing extended individualized evaluations.

e) Ease of interpretation.

Objective tests that report a single score are usually easiest to interpret, and individualized written comments are more difficult to interpret. You can share to students key and other materials that provide meaning to different scores or grades.

f) Cost.

Like other practical aspects, it is best to use the most economical assessment. However, it would be certainly unwise to use a more unreliable or invalid instrument just because it costs less.

E. How and When Should Assessment Be Done

Assessment should be continuous, not restricted to final examinations. The efforts and achievements of the students should be monitored and assessed in different ways and feedback should be provided. Thus, formative feedback, which aims to point out the strengths and weaknesses of the students, is an integral part of the assessment. Additionally, continuous assessment in various forms such as quizzes, assignments or mid-term exams should be part of the assessment and evaluation process. There are varieties of methods to assess students throughout the course that include both formative and summative evaluation. Methods for assessing student achievement of course learning objectives include multiple choice, short answer questions, and structure recognition quizzes in gross anatomy, histology, and developmental biology/embryology; practical and written tests of anatomy-radiology; standardized patient and simulator interaction, triple jump examination, and assessments for problem-based learning sessions or journal clubs considering attendance, participation, and quality of contributions.

F. Formative Assessment

Formative assessment is used during the learning program in order to check and modify the teaching and learning objectives of the program, as well as giving feedback or information to the students about their progress, strengths and weaknesses. It is also called as educational or classroom assessment. Black and William (1998) give an umbrella definition of formative assessment that is "all those activities undertaken by teachers, and/or by students, which provide information to be used as feedback to modify the teaching and learning activities in which they are engaged." Along similar lines, Cowie and Bell (1999) define formative assessment as "the process used by teachers and students to recognize and respond to student learning in order to enhance that learning, during the learning". Nicol and Macfarlane-Dick (2005), who emphasise the role students can play in producing formative assessment, state that "formative assessment aids learning by generating feedback information that is of benefit to students and to teachers. Feedback on performance, in class or on assignments, enables students to restructure their understanding/skills and build more powerful ideas and capabilities."

Formative assessment includes a variety of procedures such as observation, feedback, and journaling. Key requirements for successful formative assessment include the use of quality assessment tools and the subsequent use of the information derived from these assessments to improve instruction. The defining characteristics of formative assessment are its interactive or cyclical nature (Sadler, 1988). Formative assessment can result insignificant learning gains but only when the assessment results are used to inform the instructional and learning process (Black & William, 1998).

There are different procedures of formative assessment:

1. Observation: A teacher observes and records a student's level of engagement, academic and /or affective behaviour; develops a plan of action to support that student; implements the plan; and continues to record observation to determine its effectiveness.
2. Feedback: A teacher provides written or oral feedback to student discussion or work. For example, a teacher responds orally to a question asked in class; provides a written comment in a response or provides feedback on student work.

3. Self-assessment (Peer assessment): Students can monitor their activities by themselves. With this, students are able not only to work independently but also to assess their own performance and progress.
4. Curriculum based measurements: Standardized measures can be used to judge the progress and performance of student (Deno, 2001).
5. Portfolios: A portfolio is a collection of various forms of evidence of achievement of learning outcomes. The introduction of portfolios to the health professions, particularly as an assessment tool, has been recent, but rapid, with portfolios now regularly used at undergraduate and other educational levels. (Davis, 2001) The reason for interest in portfolio assessment is twofold: its potential to assess performance and its potential to assess outcomes, such as attitudes and professionalism, that are difficult to assess using traditional instruments.
6. Multiple Choice Questions: MCQs especially true and false type are still good tools in the formative assessment because it covers a large content area relatively written easily, but nowadays one best type of MCQs are a popular method of assessing medical knowledge because of its high validity and reliability, ease of making answer scripts, usefulness as an aid to learning and self- assessment, and high objectivity (Nnodium 1992).

While searching the formative assessment a consistent finding on formative assessment is its interactive nature which leads to significant learning gains (Black & William, 1998; Herman et al., 2006). Reviews of research on formative assessment processes support the use of questioning, observation, and self-assessment. Similarly, research has demonstrated positive effects on student achievement with the use of curriculum-based measurements (Stecker et al., 2005).

Advantages of formative assessment:

1. Occurs during instruction to provide information needed to adjust the teaching and learning objectives.
2. Frequently on daily or weekly basis.
3. More flexible according to needs of students.
4. Students play an active role with honesty, as it is non-graded.
5. Clearly defines the learning targets.
6. Provides immediate feedback.
7. Cuts down teacher preparation and grading time.
8. Provides opportunities for teacher to evaluate their own performance.

Disadvantages of formative assessment:

1. Despite its many benefits it is time consuming.
2. Serious and prolonged dedication is required.
3. It is laborious.
4. It is difficult to obtain good feedback on large scale
5. Logistic limitations are also present.

G. Summative Assessment

While summative assessment provides professional self-directive and responsibility or accountability, it also organizes standardized criteria to achieve targets by acting as a barrier to further practice or training. (Harlen 2005; Sullivan 2005). In the other words summative assessment is characterized as assessment of learning and is contrasted with formative assessment, which is assessment for learning (Glickman et al., 2009). By definition the formative and summative assessment seem diverse types of assessment; actually these are not, but are considered separately just because both are dealing with different purposes; in actual sense the same information, assembled in the same way, would be called formative if it were utilized to help learning and teaching, or summative if it were only employed for recording and reporting (Harlen 2005). Briefly, 'summative assessment' is a test that usually given at the end of a term, semester or year, the purpose is evaluative (Swearingen 2002; Garrison and Ehringhaus 1995). It is used to assess students learning and growth after instruction, whether long term learning goals have been achieved or not. (The National Centre for Fair and Open Testing 1999). It also provides a concise summary of a student's abilities in the form of pass/fail or grades A-E, 1st, 2nd, 3rd etc. (Schuwirth and van der Vleuten 2004).

Summative assessment is commonly used to refer to assessment of educational faculty by their respective supervisor with the purpose of measuring all teachers on the same criteria to determine the level of their performance. (Glickman et al., 2009). Summative assessment communicates a clear cut message both to the teachers and the students that what areas of their learning and teaching are strong and which areas would require more efforts. In other words the summative assessment is characterized to provide a procedure to assess or grade educators' level of learning in certain period of time by using well defined evaluation methodologies (i.e. fixed time and content). It provides descriptive analysis (i.e. in order to give a grade, all the activities done throughout the year are taken into account). Positively tending to stress upon students' ability as what can do rather than what they cannot (Glickman et al., 2009). The results of a summative assessment need to be compared with some sort of a standard; this could be within the class, city, province or national standard (Glickman et al., 2009).

The Impact of High-Stakes Summative Assessment on Students' Motivation

The impact of summative assessment on students' motivation for learning can be both direct and indirect (Harlen, 2005). A direct impact can be through enticing test anxiety and apprehension among the students especially the female students. The negative effect on low-achievers is more pronounced; low-achieving students had lower self-esteem than higher-achievers, whereas before exams there had been no association exists between self- esteem and success. Hence gap between high- and low-achievers' motivation is broadened. However tests do motivate some students but frequent test exercises strengthen low self-esteem of low achievers. Since it declares result in fail and pass, therefore level of anxiety is higher among students, especially female students (http://eppi.ioe.ac.uk/cms/Default.aspx?tabid=617; Harlen, 2005).

An indirect impact can be through the effect on their teachers and the curriculum. Any negative impact on motivation for learning is clearly highly undesirable, particularly at a time when the importance of learning to learn and lifelong learning is widely hold close. Thus the process of driving up test scores could have serious consequences for the current generation of students. This prompted a systematic review for evidence on the impact of summative assessment and testing on motivation for learning. (Harlen, 2005). Summative assessments can be limiting for the most intelligent students. It promotes "extrinsic" motivation, in which examinee respond to the promise of some kind of reward rather than "intrinsic" motivation in which they perform because they are interested and want to do the work" (http://eppi.ioe.ac.uk/cms/Default.aspx?tabid=617).

It is generally accepted that assessment must be fair and defensible, reliable and valid and should encourage the desire to deep learning of appropriate domains of knowledge, skills and attitudes. Additionally it should have room for individual differences and, should not disadvantage any particular group of learners (Kelly and Dennick 2009).

Anatomy provides a detailed knowledge of human structure in different forms: Gross anatomy that deals with the anatomical structures that can be dissected or seen with the naked eye; microscopic anatomy deals with minute anatomical structures on a microscopic scale; developmental anatomy studies structural changes of an individual from fertilization to maturity; neuroanatomy studies nervous tissue and the nervous system (Gray 1918). Therefore, the discipline of anatomy has a very central role in the process of training doctors and supporting medical practice since centuries (Inuva et al., 2012). A single examination to assess the level of ample anatomical knowledge attained by students does not fulfil the functions of assessment including knowledge, comprehension, skills, motivation, and feedback (Abraham et al., 2009; Lowry, 1993; Seale et al., 2000). Hence a good comprehensive test which measures multiple domains including theoretical, practical, and clinical knowledge is necessary. That tool should assess problem solving skills with more conceptualisation, more reasoning, aspects useful for meaningful reflection of its application and less memorising or rote learning, (Inuva et al., 2012; Pandey and Zimitat 2007; Tooth et al., 1989).

To test student's level of competence, summative assessments are often created in the form of:

1. written test to check theoretical knowledge, including MCQs, Clinical scenario based MCQs, Extended-matching items and SAQs
2. a performance test such as OSPE / OSCE or oral examination to check practical knowledge.

Assessment tools selected must be valid, reliable, and practical and have an appropriate impact on student learning. The preferred assessment tool will vary with the outcome to be assessed. (Shumway and Harden 2003).

Multiple Choice Questions (MCQs)

To evaluate theoretical knowledge of doctors and undergraduate medical students, multiple choice questions were introduced into medical examinations in 1950s (Al-Rukban

2006). Since then these have been proven to be more reliable tool of assessing students' knowledge rather than the traditional long essay questions (Al-Rukban 2006; Edward 2001). Since MCQs offer a large number of examination items that can cover many content areas, and allow students to choose one of the best answer from a list of possible answers or distracters, these become an ideal tool to discriminate students' ability to recall or apply accurate information. These are purposefully designed in depth of knowledge and comprehensiveness to create situations in which an examinee can answer a question by recognizing the correct option, but in absence of these options one could be unable to answer it, thus discourage students to prepare for the examination by cramming and rote memorizing of content. (Shumway and Harden 2003; Schuwirth et al., 1996; 2004).

Like other methods of assessment, MCQs also have some weaknesses and strengths. Scoring of the questions is easy, reliable, efficient, objective and can be computerized. Hence large groups of students can be assessed in a relatively short period of time. Large body of questions can be asked, while covering a broad knowledge domain. (Premadasa 1993; Edward 2001, Elstein1993; Lowe1991; Mccoubrie 2004) Comprehensively-designed MCQs can judge higher-order cognitive processing such as interpretation, synthesis and application of knowledge rather than the test of recall of isolated facts (Case and Swanson 2001). Though writing the questions requires considerable effort, their high objectivity makes it possible for the results to be released immediately after marking by anyone including a computer (Premadasa 1993; Lowe1991). Besides their strength, other concerns regarding the validity of the test: cueing effects present serious limitations, as well as the fact that recognition is tested, when sometimes recall would be desirable (Rademakers et al., 2005). It is generally emphasized that MCQs should not be used as an exclusive tool of assessment in summative examinations, but other test forms must be added in the test format (Premadasa 1993; Lowe1991).

Clinical Scenario Based Multiple Choice Questions

This presents another good option to test students' clinically oriented theoretical knowledge; by adding some relevant information to the case scenario students are asked to assess the degree to which this new information increases or decreases the probability of a particular outcome (Farmer 2005 and Charlin et al., 2000). Therefore, clinical scenario based simulated items may provide insight into clinical judgement in the real world. Answers to such types of question can easily be correlated with the students' ability and level of training and to predict future performance or oral examination of clinical reasoning (Brailovsky et al., 2001).

Extended-Matching Items (EMIs)

Are actually forms of multiple choice questions these are arranged into sets by using a list of options/ choices for all items in the set. The structural organization of such questions includes a subject or a theme / idea, a choice list, an introductory statement and at least two item stems. A typical set of EMIs begins with an option list of four to 26 options; generally more than ten options are used. (Case and Swanson 1993; Beullens et al., 2002; Al-Rukban

200). The option list is followed by two or more patient-based items and students are required to indicate a clinical decision for each item. The examinee is asked to match one or more choices mentioned in each item stem (Al-Rukban 2006). Unlike MCQs EMIs are more difficult, more reliable, more discriminating, and capable of reducing the testing time. In addition, they are quicker and easier to write. (Case 1993; Beullens et al., 2002). Since they can test diagnostic ability and clinical judgment, therefore these become a standardized criterion for clinical assessment (Premadasa 1993).

Short Answer Questions (SAQ)

Short answer questions involve more complex cognitive processes and allow for more centralized answers. In the presence of clear grading guidelines, the structured short answer question can be psychometrically excellent (Epstein 2007). SAQ are generally created around a theme based on clinical scenario followed by several pertinent and focused questions. These questions may include key features, comparisons, mechanism of actions, side effects etc. Each question bears separate mark that is clearly indicated in the question paper (Amin and Eng 2003). The total duration of time for answering the question is usually comparable or shorter than traditional long essay questions (Amin and Eng 2003). Through short answer questions a broader content areas can be covered by asking several distinct and significant questions about the topic to improve content validity. Short and specific answers made scoring very easier and better.

The reliability is improved with standard predetermined answers for each question set by the examiners. Because of these advantage SAQs become more attractive to both the examiners and the students. Like other methods of assessment SAQs have strengths and weaknesses. Short answer questions have the advantage of avoiding cueing and requiring students to supply an answer, rather than selecting or guessing from options supplied. Answers are easier to mark. The form of the questions is familiar to students and they feel less anxious in examination situations. However complex issues cannot always be satisfactorily addressed in short answers (Bacon 2003; Newble and Cannon 1989).

Objective Structured Practical Examination (OSPE)

OSPE is a variant of OSCE or 'Objective Structured Clinical Examination' initially described by Harden and his colleagues in Dundee University and subsequently developed by others as an integral part of clinical examinations (Qureshi et al., 2012; Harden and Gleeson 1979; Harden et al., 1975). Students have a psychology to pass the examination; therefore their entire interest is to know what material is exactly being covered in the examination so to concentrate on that rather to learn too much. If the practical exam should be made comprehensive by including clinical problems, with a wide range of skills and practical materials covered in lab sessions in an objective fashion, students will surely study each clinical problem, skills and practical material in such a way that they understand it well rather than a rote memorization. Therefore an OSPE / OSCE is a good tool for the assessment of students practical knowledge. Students will surely spend time with lab materials / patients to make sure they have mastered all the skills if they know there is going to be an OSPE / OSCE (Prozesky 2001).

Because of being comprehensive, and reliable, the objective structured practical examination (OSPE) is now an accepted tool in the assessment of practical skills in pre- and para-clinical subjects. This is truly a structured examination into which a variety of test methods can be incorporated to test a wide range of skills in an objective fashion. (Nageshkumar 2005). OSPE has already entered as the final examination conducted in many disciplines including anatomy in many medical institutes / Universities. Anatomy questions generally constitute 40–60% of the OSPE stations incorporating gross, microscopic, developmental and neuroanatomical specimens integrating with pathological, radiological or other materials. Anatomy stations test skills of identification, interpretation of investigational results in the anatomical context, surface markings and living anatomy. The examination usually comprises of 25–40 stations, depending upon the number of students and the nature of the skills-components. Each station is given the same time duration and utilizes anatomical specimens including skeletal parts / bones, plastic models, prosected wet cadaveric and plastinated specimens, histological slides, and imaging films (radiographs, CT scans, MRIs) with clinical correlation and application (Chakravarty et al. 2005).

The Oral Examination

For its flexibility and its potential for testing higher order cognitive skills oral examination was included as an important component of the final summative examination in many medical schools since centuries to test the practical knowledge of students (Anderson et al., 1965; Wakeford et al., 1995). Face-to face interaction between the examiner and examinee was considered to provide a unique chance to test communicating skills, and this dialectic communication was thought to be indorsed very valuable training for future career interviews; which otherwise cannot be assessed by any other examining tool. Gradually it was felt that the oral exam is losing its validity and reliability due to various factors: (a) many students are anxious before the oral examination (Glenn Pearce and Geoffrey Lee, 2009; Newble et al., 1987), which could actually disturb students "thinking process" and overall performance; (b) there is considerable wide range of variations of questions from student to student, thus some students may get easy questions other gets difficult questions; (c) inconsistency of the examiners in the form of general tendency toward leniency, and errors of contrast, and various other circumstances including examiners mostly indulge in over-marking in viva-voce examinations in order to make an otherwise undeserving candidate 'pass' (Baral 2009; Torke et al., 2010; Ghosh et al., 2012). Another disadvantage that has largely been observed in oral examination is that confidence and good verbal skills are considered to be the most favourable objects for the students to get good score, therefore comparatively less knowledgeable, weaker but stress-free, confident, and loquacious students may get a better score as compared to a hardworking, knowledgeable student who has difficulty in expressing him/herself. (Baral 2009; Torke et al., 2010; Thomas et al., 1993). The other common deficiencies of orals are that the questions are rarely prepared in detail beforehand, but are posed extemporaneously as the examiner is prompted by the student's responses. With these limitations, oral examination cannot be considered as a good tool for assessment. It is desirable that the dependency on oral examination must be considerably reduced. However, to test higher order cognitive skills to discriminate top students, structured, standardized orals using hand-picked examiners could be possible option (Torke 2010).

Conclusion

Anatomy, being one of the core disciplines in medicine deserves great attention both for teaching-learning and assessment-evaluation strategies. A good assessment and evaluation system should include a multitude of methods and assessment at different time points. Formative and summative assessments should be carried out regardless of the type of curriculum, i.e., whether it is a discipline-based traditional or integrated and problem-based curriculum. As most medical schools now have integrated curricula, either with or without PBL, anatomy is learned and taught in clinical context. Therefore, it is a necessity that assessment and evaluation should also contain clinically relevant components. While OSPE and OSCE are excellent tools to accomplish this, MCQs with clinical vignettes should be encouraged. Formative assessment in various forms should be an integral part of assessment system to provide the students their strengths and weaknesses. A good formative assessment also provides a good feedback to curriculum planners and course organizers. The weightage of continuous assessment can be planned according to the type and content of the course(s) in the curriculum. Summative evaluation can use different evaluation tools, mainly OSPE, OSCE and MCQs. When planning the assessment-evaluation system, psychometric properties such as validity, reliability, fairness should be observed. Due to the subjectivity and unfairness, oral tests are no longer among the preferred methods.

References

Al-Rukban M O (2006) Guidelines for the construction of multiple choice questions tests *Practical Tips* 13 (3) 125-133.

Amin Z, Eng K H (2003) Basics in Medical Education. World Sci publishing Co pte Ltd.

Anderson J, Roberts FJ (1965) A new look at medical education. London: Pitman.

Armstrong J, Conrad L. (1995) Subject review and evaluation: a resource book for improving learning and teaching. Nathan, Queensland: Griffith Institute for Higher Education.

Bacon D R (2003) Assessing Learning Outcomes: A Comparison of Multiple-Choice and Short-Answer Questions in a Marketing Context *Journal of Marketing Education* 25 (1) 31-36.

Baral P (2009) How to Improve the Viva Voce: Some Guidelines. Alert Bulletin 2 (3): 60-61.

Beullens J, Van Damme B, Jaspaert H, Janssen PJ.(2002) Are extended-matching multiple-choice items appropriate for a final test in medical education. *Med Teach*; 24(4):390-5.

Biggs J (2001) Assessing for quality in learning. In Suskie L (ed.)Assessment to promote deep learning. Insight from AAHE's 2000 and 1999 Assessment conferences. USA: American Association for higher Education.

Black, P., & William, D. (1998). Assessment and classroom learning. *Assessment in Education*, 5(1), 7-74.

Brailovsky C, Charlin B, Beausoleil S, Cote S, Van der Vleuten C. (2001) Measurement of clinical reflective capacity early in training as a predictor of clinical reasoning performance at the end of residency: an experimental study on the script concordance test. *Med. Educ.*; 35:430-6.

Brookhart, S. M. (1999), Teaching About Communicating Assessment Results and Grading. 18 (1) 5–13.

Case SM, Swanson DB. (1993) Extended matching items: a practical alternative to free response questions. *Teach Learn Med.*; 5:107-15.

Case SM, Swanson DB. (2001) Constructing Written Test Questions for the Basic and Clinical Sciences, 3rd ed. Philadelphia: National Board of Medical Examiners.

Chakravarty M, Latif NA, Abu-Hijleh MF, Osman M, Dharap AS, Ganguly PK. 2005 Assessment of anatomy in a problem-based medical curriculum. *Clin. Anat.* 2:131-6.

Charlin B, Roy L, Brailovsky C, Goulet F, van der Vleuten C. (2000) The Script Concordance test: a tool to assess the reflective clinician. *Teach Learn Med*; 12:189-95.

Edward M. Multiple choice questions: their value as an assessment tool. Lippincott Williams & Wilkins, Inc, 2001. Volume 14(6), pp 661-666.

Cowie B, Bell B (1999). "A model of formative assessment in science education". *Assessment in Education* 6: 101-116.

Davis MH, Friedman Ben-David M, Harden RM, Howie J, Ker J, McGhee C, Pippard MJ,Snadden D. (2001) Portfolio assessment in medical students final examination. *Med. Teach* 23:357-366.

Deno, S. L. (2001). Curriculum-based measures: Development and perspectives. Retrieved April 18, 2008, from htpp://www.progressmonitoring.net/CBM Article Deno.pdf.

E How educational learning http://www.ehow.com/info_8502289_advantages-diasadvatages_formative assessment.html

Elstein AS (1993). Beyond multiple-choice questions and essays: the need for a new way to assess clinical competence. *Acad. Med.*; 68:244-9.

Epstein R M (2007) Assessment in Medical Education. N Engl J Med 356:387-96.

Farmer EA, Page G. A (2005) practical guide to assessing clinical decision-making skills using the key features approach. *Med. Educ.*; 39:1188-94.

Garrison, C. & Ehringhaus, M. (1995) "Formative and Summative Assessments in the Classroom."http://www.nmsa.org/Publications/WebExclusive/Assessment/tabid/1120/Default.aspx

Ghosh A, Mandal A, Das N, Tripathi S K., Biswas A, and Bera T (2012) VIVA VOCE Indian *J. Pharmacol.* Mar-Apr; 44(2): 274–275.

Glickman, C.D., Gordon, S.P., & Ross-Gordon, J.M. (2009).Supervision and instructional leadership: a developmental approach Allyn and Bacon, Boston, MA.

Case SM, Swanson DB. (2001) Constructing Written Test Questions for the Basic and Clinical Sciences, 3rd ed. Philadelphia: National Board of Medical Examiners.

Fuchs, L. S., D. Hosp, M. K., Jenkins, J. R. (2001). Oral reading fluency as an indicator of reading competence: A theoretical, empirical, and historical analysis. *Scientific Studies of Reading*, 5, 239-256.

Gray H (2000) Anatomy of the Human Body. Introduction page, 20th edition. Philadelphia: Lea & Febiger, 1918 New York: bartleby.com.

Harden RM, Stevenson M, Wilson DW, Wilson GM (1975) Assessment of clinical competencies using objective structured clinical examination. *BMJ* 1: 447-451.

Harden RM, Gleeson FA (1979) Assessment of clinical competencies using an objective structured clinical examination (OSCE). *Med. Educ.* 13:41-54.

Harlen W (2005) Teachers' summative practices and assessment for learning – tensions and synergies *The Curriculum Journal*, 16 (2): 207 – 223.

Headington R, Kerrigan-Holt M (2011) .Holistic assessment design: Reality, possibility or elusive ideal? Assessment in Higher Education Conference, 6 July, University of Cumbria.

Herman, J. L., Osmundson, E., Ayala, C., Schneider, S., & Timms, M. (2006). The nature and impact of teachers formative assessment practices. CSE Technical Report #703. National center for Reasearch on evaluation, Standards, and Student Testing (CRESST). http://www.provost.cmich.edu/assessmenttoolkit/formativesummative.htm.

Hintze JM, Theodore J. Christ, Scott A. Methe (2006) Curriculum based assessment. Psychology in the school, vol. 43(1). DOI:10.1002/Pits.20128.

Inuwa IM, Taranikanti V, Al-Rawahy M, Roychoudhry S, Habbal O. (2012). "Between a Rock and a Hard Place": The discordant views among medical teachers about anatomy content in the undergraduate medical curriculum. *Sultan Qaboos Univ. Med. J.* 1:19-24.

Kellough, R.D. and Kellough, N.G. (1999). Secondary School Teaching: A Guide To Methods And Resources; Planning For Competence. Copyright by Prentice Hill, Upper Saddle River, New Jersey.

Kelly S and Dennick R (2009) Evidence of gender bias in True-False-Abstain medical Examinations BMC *Medical Education*, 9:32.

Khan J S, John SGB, Saima T (2010) Assessment in medical education in the 21st century. *J. Ayub Med Coll Abbottabad*;22(3).

Le Roux K (2012). A cross-sectional survey to compare the competence of learners registered for the Baccalaureus Curationis programme using different learning approaches at the University of the Western Cape, Curationis 34(1):53-60. http://dx.doi.org/10.4102/Curationis.34i1.53

Lesky E. Medical Education in England since 1600, In O'Malley CD. (Ed.) The history of medical education. London: University of California Press; 1970.p.235–50.

Lowe D (1991) Set a multiple choice question (MCQ) Examination. *BMJ*; 302(6779):780-2.

Lowry S. (1993) Assessment of students. *Br. Med. J.* 306: 51–54.

Mccoubrie P (2004). Improving the fairness of multiple-choice questions: a literature review. Med Teach; 26(8); 709-12.

McMullan M, Edacott R, Gray MA et al. . Portfolios and assessment of competence: a review of literature. http://cms.kcn.unima.mw:8002/moodle/downloads/chilemba/My%20Documents/Gladys/Portfolios%20and%20assessment%20of%20competence.pdf

Miller GE (1990) The Assessment of clinical skills /competence /performance; Acad .Med:65 (9); 63-67. http://www.gptraining.net/training/educational_theory/adult_learning/miller.htm

Nageshkumar G Rao. 2005 OSPE - A Proposal for Innovative Changes in UG Practical Forensic Medicine Examination Medico-Legal Update 5 (1) 005-01 - -03).

The National Center for Fair and Open Testing (1999). The Value of Formative assessment. Fair Test Examiner http://www.fairtest.org/examarts/winter99/k-forma3.html

Nendaz A, Tekian K. (1999). Assessment in problem-based learning medical schools: a literature review. *Teach Learn Med.* 11:232-243.ss

Newble D, Cannon R. (1987) A handbook for medical teachers. 2nd edn. Lancaster: MTP Press.

Newble D. Assessment. In: Jolly B, Rees L. (Editors) Medical education in the millennium, Oxford: Oxford University Press; 1998.p.131–42.

Newble, D. and Cannon, R.. A Handbook for Teachers in University and Colleges. New York: Kogan 1989 Page, p 107.

Nicol D, Macfarlane-Dick D (2005). Rethinking Formative assessment in HE: a theoretical model and seven principles of good feedback practice. *Quality Assurance Agency for Higher Education*.

Nnodium JO (1992) Multiple choice testing in Anatomy. *Med. Educ.* 26:301-9.

Norcini JJ (2003) Work based assessment. *BMJ.* April 5; 326 (7392): 753–755. doi: 10.1136/bmj.326.7392.753

Nygaard C, Holtham C, Courtney N.(2009) Improving Students' Learning Outcomes edited by Copenhagen Bussiness school press.1st ed. ISBN 978-87-630-0232-5.

Palomba, C.A. & Banta, T.W. (1999) Assessment Essentials: Planning, Implementing, and Improving Assessment in Higher Education. San Francisco: Jossey-Bass, p. 4.

Pandey P, Zimitat C. (2007) Medical students' learning of anatomy: memorization, understanding, and visualization. *Med. Educ.*; 41:7-14.

Parker N (2001) Student learning as information behaviour: exploring Assessment Task Processes. Information Research, 6(2). Available at: http://InformationR.net/ir/6-2/ws5.html

Pearce G and Lee G. (2009) Viva Voce (Oral Examination) as an Assessment Method: Insights From Marketing Students *Journal of Marketing Education.*

Premadasa IG (1993). A reappraisal of the use of multiple choice questions. *Medical Teacher;* 15(2-3):237-42.

Prozesky D. (2001) Assessment of Learning. *Community Eye Health* 14(38): 27–28.

Qureshi G S, Parveen S, Yasmeen T (2012) Objective Structured Practical Examination (OSPE): an experience at Dow University of Health Sciences (DUHS) Karachi. MC 18 (1) 5 – 8.

Rademakers J, Ten Cate Th J, Bar P R. (2005) Progress testing with short answer questions *Medical Teacher* 27(7):578–582.

Reem R A, Rao R, Kamath S and Kamath A. (2009). A trial of the objective structured practical examination in physiology at Melaka Manipal Medical College, India. Adv Physiol Educ 33: 21–23.

Rees L, Jolly B. (1998) Medical education into the next century. *Medical Education in the millennium*. Oxford 245–60.

Ronald JC and Swerdlik ME (2004).Psychological Testing and Assessment : An Introduction to Tests and Measurement, ISBN-13: 9780072887679.

Sadler, D. R. (1988). Formative assessment: Revisiting the territory. *Assessment in Education*, 5, 77-84.

Schuwirth LW, van der Vleuten CP, Donkers HH (1996) A closer look at cueing effects in multiple-choice questions. Med. Educ.; 30:44-9.

Schuwirth LW, van der Vleuten CP (2004). Different written assessment methods: what can be said about their strengths and weaknesses? *Med. Educ.*;38:974-9.

Scriven, Michael, C. (1967). The methodology of evaluation. In Gredler, M. E. Program Evaluation. (p. 16) New Jersey: Prentice Hall, 1996.

Seale KJ, Chapman J, Davey C. (2000) The influence of assessments on students' motivation to learn in a therapy degree course. *Med. Educ.* 34: 614–621.

Shumway J.M. & Harden R.M. AMEE Guide No. 25: (2003) The assessment of learning outcomes for the competent and reflective physician *Medical Teacher*, Vol. 25, No. 6, pp. 569–584.

Stecker, P. M., Fuchs, L. S., & Fuchs, D. (2005). Using curriculum-based measurements to improve student achievement: Review of research. *Psychology in the Schools*, 42, 795-819.

Sullivan W. Work and integrity: the crisis and promise of professionalism in America. 2nd ed. San Francisco: Jossey- Bass, 2005.

Summative assessment http://eppi.ioe.ac.uk/cms/Default.aspx?tabid=617

Swearingen R (2002) A Primer: Diagnostic, Formative, & Summative Assessment, Heritage University, Thomas CS, Mellsop G, Callender J, Crawshaw J, Ellis PM, Hall A, et al. (1993). The oral examination: a study of academic and non-academic factors. *Med. Educ.* 27: 433-439.

Tooth D, Tonge K, McManus IC. (1989) Anxiety and study methods in preclinical students: causal relation to examination performance. *Med. Educ.* 23: 416–421,

Torke S, Abraham RR, Ramnarayan K, Asha K. (2010) The impact of viva-voce-voce examination on students' performance in theory component of the final summative examination in physiology. *J Physiol Pathophysiol* 1:10–2.

Wakeford R, Southgate L, Wass V (1995) Improving oral examinations: selecting, training, and monitoring examiners for the MRCGP.*BMJ* 311: 931-935.

Zhao J. (2011) Student evaluation of instruction in higher education: exploring issues of validity and reliability, *Assessment & Evaluation in Higher Education*.37(2) ;239-249.

In: Education in Anatomical Sciences
Editor: Paul Ganguly
ISBN: 978-1-62618-488-6

Chapter 15

WRITING GOOD MCQS

Raja G. Subhiyah and M. Brownell Anderson*
National Board of Medical Examiners,
Philadelphia, US

ABSTRACT

This chapter is intended largely for the busy anatomist who is giving a course to medical students. Part of the activity involved in a course is, of course, assessing the work of professor and students alike. We believe that preparing the assessments goes hand in hand with preparing the course materials. \It all starts with the desired learning outcomes, and both teaching/learning materials/methods, and the means of assessing them, become two faces of the same goal: effective learning.

Although assessing the outcomes of the teaching/learning experience involves a multi-faceted operation, Multiple Choice Questions (MCQs) form one of the cornerstones of this process. It is therefore our intent to concentrate on that cornerstone of assessing the outcomes of the course. This effort is intended as a practical guide to producing good MCQs. We will not go into theories and conjecture, nor will we assume complete inexperience with writing and using MCQs. Since we assume that the reader is a busy professor with some previous experience in writing MCQs, we have tried to offer a direct, succinct, and practical guide. We have supplied what we view as necessary to write really good MCQs.

INTRODUCTION TO ASSESSMENT

The Purposes of Assessment

In this chapter, we will limit our discussions to MCQ tests used in courses and/or teaching units in, say, a university setting. The purpose of assessment is to give an estimate of proficiency in a certain area of knowledge. More specifically, in a university or school

* E-mail: mbanderson@nbme.org.

setting, assessment ascertains the extent to which learning objectives outlined in the curriculum are being met. This assessment may be used for two generally accepted purposes:

1. Summative evaluation: the extent to which objectives of a course or a teaching unit have been achieved.
2. Formative evaluation: determine which objectives in a teaching unit or course need further development and/or remedial action.

These uses for assessment involve the discovery of weaknesses and strengths in student performance, teaching performance, and curriculum.

The Alignment of Assessment and Curriculum

In a university or a school setting, the relevance and utility of assessment is directly linked to the curriculum. These two processes, assessment and application of curriculum, often influence one another. Thus, assessment is driven and guided by the contents of the curriculum, while the contents of the curriculum may be modified in light of results of assessment. It is therefore very important to establish a close correspondence between teaching/learning objectives and assessment. Another way to consider this is, how do we determine if the educational/behavioral outcomes we have established are being achieved?

In fact, we strongly suggest that questions for the assessment be written concomitantly with preparation of teaching materials. Thus, part of a lesson plan would include the questions that are going to be used for measuring its success. One likely sequence of activity in preparing a teaching unit could be:

1. List the outcomes desired from the teaching unit. State the outcomes in terms of actions and be specific. E.g., students should identify the vagal nerve in a frontal chest x-ray.
2. Describe the method of achieving this (or a group of related) objective and the setting. E.g., Lecture in the classroom with three demonstration slides.
3. Write the questions that will be used to test whether students achieved the objective for use later in a summative or formative evaluation.

What's in a Score?

The primary result of an assessment is a score or a set of scores. The most important aspect of a score is how it is used, viz., the interpretation it is given and the inference that is drawn from it (validity). The validity of a score is a measure of how appropriate are the conclusions drawn from it or the decisions based on it. In assessing a course, the main form of validity is related to the extent to which the questions on the test faithfully reflect the desired learning outcomes.

The other important attribute of a score is its precision (reliability). Reliability is a measure of how well the content assessed by the test is sampled.

The test is a specific sample of items from a content domain (e.g., the nervous system). If the test adequately samples the content domain, then performance on the test caninfer the degree of proficiency in the whole domain. Such inferences, when drawn, involve error of measurement, the converse of reliability. The best way to increase reliability of an assessment is by increasing the sampling: i.e., by increasing the number of questions. This is where MCQ tests have a clear advantage, as they can contain a large number of questions and yield reliable measures.

Anatomy of an MCQ

The most important property of a good MCQ is to access the targeted concepts directly, simply, and quickly. While there are several formats for MCQs, we will concentrate our discussion on the conventional (A-type) format as research has shown that, overall, they have the most advantages. Generally, a conventional MCQ has three basic components: a stem, a correct (or best) response, and distractors.

Stems

The stem should present all the necessary information to correctly respond to the question. Some stems contain extraneous and/or contextual information, but there is no proof that by including such information the concept is better tapped.

Of course, if part of the concept being measured is to discern relevant from irrelevant information, then such additional information in the stem may be tolerated. However, in most cases, direct stems with a minimum of words works best.

The Correct or Best Response (Key)

The key could present either the only correct response or the best response. In some items, all the choices are clearly incorrect except for the keyed response.

This tends to occur when measuring simpler concepts that lend themselves to black-white thinking (e.g., the name of a vessel). When the concept being tapped is more complex, all the choices may be partially correct, but the key must be distinctly more appropriate than any of the other choices. In the conventional MCQ, only one key is entertained.

Distractors

The distractors are alternative responses that are incorrect or inferior to the key. The number of distractors varies, but most commonly MCQs have three or four distractors.

This makes the total number of alternatives four or five. A test mixing four- and five-choice MCQs is perfectly acceptable.

At the medical school level, examinees are not very likely to guess except perhaps on very difficult items, so correction for guessing is generally not recommended.

We believe that a mixture of four- and five-choice items is optimal for an anatomy test. Three-choice items may be used if it is very hard to find good distractors, or even six- or-seven-choice items. The latter may have decreased efficiency, as they take up longer testing time.

PHYSIOLOGY OF AN MCQ

The Pros and Cons of MCQs

Like any instrument, a MCQ has areas of strength and weakness. These areas depend at least in part on how the MCQ is wielded and on the context in which it is used. While an obvious area of excellence is the measurement of factual knowledge, the well-written MCQ can also measure very complex reasoning. The weakness of MCQs lies in directly measuring skills in the performance of tasks. Here are a few advantages gained from using MCQs:

1. A wide range of knowledge-based concepts can be tapped.
2. They are efficient. The information yielded about the examinee's proficiency per hour of testing time is high.
3. They are relatively easy to produce and use. It takes some skill to write good MCQs, but the skill is not very hard to master.
4. MCQs are relatively inexpensive to administer.
5. Amassed, MCQs yield reliable and highly reproducible measures.
6. Quick and easy to score.

And following are a few disadvantages:

1. Limited capability of measuring skills.
2. More effort is required to write MCQs that access higher levels of thinking.
3. Simple recall items are easy to write and tend to induce the author to overburden the test with this kind of item.
4. They are somewhat prone to test-wiseness bias.

What Can MCQs Measure Best

The main arena in which MCQs can successfully be used is knowledge. Knowledge, however, is not synonymous with simple recall. It spans the gamut of thinking processes and includes comprehension of concepts, application of principles, complex deductive and inductive reasoning, as well as evaluation of situations and principles. The secret in obtaining these functions is in writing the questions.

Function of the Stem

The stem furnishes the information and lays the background for selecting one of the choices. It should contain all the necessary information to correctly respond to the item. The stem may be in the form of a question, e.g.:

What is the most cost-effective way of cooling a room in warm, dry weather?

a. Air conditioning
b. Cooling tower
c. Electric fan
d. Evaporative cooler

or it may also be in the form of an incomplete statement e.g.:

The most cost-effective way of cooling a room in warm, dry weather is

a. air conditioning
b. cooling tower
c. electric fan
d. evaporative cooler

Function of the Choices

The key should clearly furnish the best response to the problem or question posed in the stem. In the classical form of the MCQ, there should only be one such best response. Any ambiguity or dispute of the key being the best response must be avoided.

The distractors' function is to offer plausible but incorrect alternatives to the key. They are designed with the intention of successfully drawing in those who are not knowledgeable. Thus, a good distractor, without undue fuss, will seem correct to all but the examinees who know the target concept.

Example:

A 23-year-old woman comes to the office because of a slow-growing mass on her neck for the past 3 months. Examination shows a 4-cm, palpable, freely movable, non tender mass just below the angle of the mandible. Ultrasonography indicates that the mass is cystic.

Which of the following is the most likely diagnosis?

a. Branchial cyst
b. Lymphangioma
c. Parathyroid adenoma with cystic change
d. Reactive lymph node with cystic degeneration
e. Thyroglossal duct cyst

Quality of an MCQ

What Makes a Good MCQ?

Clarity. The successful MCQ be written with language that is clear in conveying information and responses. There should be no ambiguous language or blanket statements (with "all," "always," or "never"). The writer should also avoid words with widely-ranging interpretation and fuzzy meaning (e.g., "good," "bad," or "ugly").

Simplicity. The writer should impart the information and responses using the simplest appropriate language. This includes both vocabulary and grammatical structures. No artificial difficulty due to language should find its way into the MCQ. The difficulty of the MCQ should only stem from that of the concept being measured. This lack of linguistic sophistication is particularly important for the validity of the questions when the target population is testing in other than the mother tongue.

Another issue that involves simplicity is whether to add extra information to the stem in the interest of creating a realistic "scenario" as a background for the item. The vignette or scenario will add words and cost time. It may also introduce complexity and alter the concept being sought with respect to the learning outcome. However, the vignette will add "authenticity" to the item and may increase its face validity. We believe that if the scenario is a part of the learning outcome to be achieved, and is important to the learning process, by all means it should be used. If, however, the scenario is not an integral part of the learning outcome to be achieved, it should be omitted.

Omitting unnecessary scenarios has advantages such as:

- makes the items easier and less time-consuming to write,
- costs the examinee less time, so more items can be used in the same testing time

Economy. When writing MCQs, the efficiency of the overall assessment is an important consideration. Testing time is usually limited and precious. Thus, full advantage of testing time dictates that a maximum of information about the examinees' proficiency is collected. The MCQ writer must strive to make the assessment as economical in terms of time allotted as possible by avoiding unnecessary and extraneous complexities in the language and structure of the MCQ. A simple, direct approach is what is needed. Minimizing word count, employing direct and simple grammar structures, avoiding long sentences, and using the simplest appropriate vocabulary all contribute to economy of measurement.

Focus. The content of an MCQ must focus on only one concept. While it is known that some concepts are more complex than others, nevertheless, there must be only one question to ask. If there is more than one concept involved in the question, the response may well confound the inferences drawn from the item. Here, it may be useful to think of the distinction between "complex" and "complicated" while authoring a MCQ.

Connection with the concept being measured. The item must accurately reflect the intent of the assessment. In other words, the likelihood that an examinee's response to the item is correct should vary positively as his/her degree of capturing the concept. Thus, a correct response must indicate a stronger grasp of the concept than an incorrect one. This is the essence of a valid MCQ.

What Makes a Good Stem?

A good stem should effectively and efficiently lead the proficient examinees to the correct response. It should have the following attributes:

- Contains all necessary information to identify the correct response.
- Is simple and direct. No artificial complications such as unnecessary information, complicated language structures, vocabulary, and/or complicated formats.
- Short. Should have the minimum necessary word count.
- Focused on only one concept.
- Directly and strongly linked to the concept being tapped.
- Directly and strongly linked to a learning outcome (specific behavior).

What Makes Good Choices?

The choices should follow the same style as the stem. They should be simple, direct and clear. Here are a few attributes of good choices:

- Only one clearly correct response.
- The choices should all have the same style and length. The correct choice should be indistinguishable from the others except in content.
- The correct responses across items should be evenly spread among the choices.
- It may be a good idea to alphabetize the choices.
- The choices should all appear to be logical consequences of the stem.

The Art of Writing an MCQ

Writing a good MCQ is the art of transforming a collection of words into a stimulus that will accurately elicit a response from the examinee. This interaction between MCQ and respondent is laden with information, hopefully valid information, about the respondent's proficiency. An MCQ is only as good as the extent to which it yields the right information the tester is seeking. Thus, the author must keep clear the objective of that item, and remove extraneous factors that may cloud the distillation process.

Rules of Thumb

To help arrive at the desired endpoint: producing the good MCQ, here are a few final suggestions:

- Align the question as closely as possible with a specific learning outcome. This includes the level at which you are testing (know, apply, deduce, infer, etc.).
- Ask (or elicit) only one concept per MCQ.

- Make it short and direct.
- Use the simplest language consistent with the technical jargon. Remember, this is a test in anatomy and not in English.
- Make the choices similar in format, structure, grammar, and length.
- Avoid tricks and unrelated information.
- Make sure that there is one and only one best response.

REFERENCES AND FURTHER READING

Ackerman, T.A., and Smith, P.L. (1988).A comparison of the information provided by essay, multiple-choice, and free-response writing tests. *Applied Psychological Measurement,* 12, 117-128.

Albanese, M.A., Kent, T.A., and Whitney, D.R. (1977). A comparison of the difficulty, reliability, and validity of complex multiple-choice, multiple response, and multiple true-false items. *Annual Conference on Research in Medical Education,* 16, 105-110.

American Psychological Association, American Educational Research Association, National Council on Measurement in Education. (1985). *Standards for Educational and Psychological Testing.* Washington, DC: American Psychological Association.

Birenbaum, M., and Tatsuoka, K.K. (1987). Open-ended versus multiple-choice response formats-it does make a difference for diagnostic purposes. *Applied Psychological Measurement,* 11, 385-395.

Case, S. and. (1989). *Performance of various multiple-choice item types on medical specialties examinations: Types A, B, C, K, and X.* Philadelphia, PA:: National Board of Medical Examiners.

Changas, P., and Samejima, F. (1984, April).*Efficient use of distractors in ability estimation with the multiple-choice test item.* Paper presented at the annual meeting of the American Educational Research Association, New Orleans.

Dawson-Saunders, B., Nungester, R.J., and Downing, S.M. (1989).*A comparison of single best answer multiple-choice items (A-type) and complex multiple-choice (K-type).*Philadelphia: National Board of Medical Examiners.

Downing, S. and Haladyna.

Ebel, R.L. (1951). Writing the test item. In E.F. Lindquist (Ed.), *Educational Measurement* (1st ed., pp. 185-249). Washington, DC: American Council on Education.

Ebel, R.L. (1981, April). *Some advantages of alternate-choice test items.* Paper presented at the annual meeting of the National Council on Measurement in Education, Los Angeles.

Gagne, R.M. (1968). Learning hierarchies. *Educational Psychologist,* 6, 1-9.

Haladyna, T.M. (1992b). The effectiveness of several multiple-choice formats. *Applied Measurement in Education,* 5, 73-88.

Haladyna, Thomas M. (1994). *Developing and validating multiple-choice test items.* Hillside, NJ: Lawrence Erlbaum Associates.

Haladyna, T.M., and Downing, S.M. (1989a). A taxonomy of multiple-choice item-writing rules. *Applied Measurement in Education,* 1, 37-50.

Haladyna, T.M., and Downing, S.M. (in press). How many options is enough for a multiple-choice test item. *Educational and Psychological Measurement.*

Jacobs, P.I., and Vandeventer, M. (1970).Information in wrong responses. *Psychological Reports,* 26, 311-315.

Miller, H., and Williams, R.G. (1973). Constructing higher level multiple choice questions covering factual content. *Educational Technology,* 13, 39-42.

Miller, H., Williams, R.G., and Haladyna, T.M. (1978).*Beyond facts: Objective ways to measure thinking.* Englewood Cliffs, NJ: Educational Technology.

Nitko, A.J. (1989). Designing tests that are integrated with instruction. In R.L. Linn (Ed.), *Educational measurement* (3rd ed., pp. 447-474). New York: American Council on Education and Macmillan.

Norcini, J.J., Swanson, D.B., Grosso, L.J., and Webster, G.D. (undated). *A comparison of knowledge, synthesis, and clinical judgment multiple-choice questions in the assessment of physician competence.* Philadelphia: American Board of Internal Medicine.

Roid, G.H., andHaladyna, T.M. (1982).*Toward a technology of test-item writing.* New York: Academic Press.

Ruch, G.M., and Charles, J.W. (1928). A comparison of five types of objective tests in elementary psychology. *Journal of Applied Psychology,* 12, 398-403.

Sanders, N.M. (1966).*Classroom questions. What kinds?* New York: Harper and Row.

Stiggins, R.J., Griswold, M.M., and Wikelund, K.R. (1989).Measuring thinking skills through classroom assessment. *Journal of Educational Measurement,* 26, 233-246.

Subhiyah, R.G. and Downing, S.M. (1993).*K-Type and A-Type items: IRT comparisons of psychometric characteristics in a certification examination.* Paper presented at the annual meeting of the National Council on Measurement in Education, Atlanta.

In: Education in Anatomical Sciences
Editor: Paul Ganguly
ISBN: 978-1-62618-488-6

Chapter 16

PRACTICAL EXAMINATIONS: OSPE, OSCE AND SPOT

Muhammad Zafar*[*]*, Ahmed Yaqinuddin, Muhammad Faisal Ikram and Paul Ganguly
College of Medicine
AlFaisal University, Riyadh, Saudi Arabia

ABSTRACT

A problem-based learning (PBL) curriculum requires students to be able to integrate basic and clinical sciences. Assessment of anatomical understanding in PBL requires several types of examinations. No single method of assessment can effectively test the three individual aspects of learning; knowledge, attitudes, and skill. Anatomical knowledge can be assessed by written and practical examinations. OSPE is a tool which is used to assess anatomy laboratory skills of students in the preclinical stage of a medical curriculum. OSPE is a new concept in practical assessment of basic medical sciences. In OSPE, multiple stations are designed and each station has a specific objective which needs to be tested. OSPE is more objective, reliable and a valid tool of assessment to assess practical aspects of anatomy in an integrated problem based curriculum. Organization of OSPE requires team work, logistics but at the same time a large number of students can be tested with standard setting in a short period of time. OSPE is still considered as the most common and popular method of assessing practical or laboratory based basic sciences in the early phase of medical curriculum. We have adopted OSPE at College of Medicine AlFaisal University and we believe our OSPE appropriately tests students' grasp of anatomical knowledge and we believe it is motivating, inspiring and interesting. OSPE should follow the curriculum and demands higher cognitive skills. Therefore, there is a need to develop objective tools to assess higher cognitive skills like clinical application of anatomical knowledge.

---

[*] E-mail: mzafar@alfaisal.edu.

INTRODUCTION

"Students work to pass, not to know… They do pass and they don't know"
(Thomas Huxley)

We believe practical is the practice of theory as well as the application of knowledge, and students must have sufficient knowledge, skill and attitude to pass any examination. For a doctor, practical skills are essential, and must be learnt and mastered. In most medical schools, almost 40-50% of the curricular time is devoted to practical sessions.

It is surprising that the assessment methods used do not ensure that the skills which are expected out of the students at the end of the course are actually developed. All assessment methods must be targeted to test proficiency of psychomotor skills.

The examination must be objective, reliable, and valid and conducted in a reasonable time. Adequate knowledge of anatomy, with strong emphasis on "understanding" rather than "knowing", and the acquisition of essential anatomical skills such as the power of observation, identification and interpretational abilities, is considered crucial for effective and safe clinical practice. The learning of anatomy by medical students is important not only for the attainment of essential, clinically-based knowledge, but also for the achievement of a range of skills otherwise difficult to obtain. The medical curriculum should have a solid clinical emphasis in the basic medical sciences. Anatomical knowledge can be assessed by written and practical examinations. Written examinations assess core knowledge of anatomy in relation to clinical or functional context.

In general, cognitive ability is tested by a written examination system constituting multiple choice questions (MCQs), extended matching questions (EMQs), and short answer questions (SAQs), problem solving abilities by simulations and skills by practical examinations e.g. OSPE

Methods of Assessment of Practical Skills and Knowledge

Practical examinations, however, test the ability of students to relate clinical and functional knowledge with three-dimensional (3D) orientation of gross structures. There are different methods for the assessment of practical knowledge; oral (viva-voce), spot or spotting, objective structured practical examination (OSPE), and objective structured clinical examination (OSCE).

Spotting or Spot Examination

"Spotting or Spot" examination constitutes a series of stations exhibiting cadaveric parts, dissections, plastinated models, microscopic slides, photomicrographs and radiological/CT images with arrows, pins and tags pointing to specific structures (popularly referred to as "spotters"). At each station, specific questions are then asked about where the pin is placed so as to identify a structure, and a question related to that structure e.g. action of a muscle, branches of an artery, relations of a viscera, functions of a cell etc. A typical spot examination comprise 20-30 stations each lasting 1 minute and each with an A and B part. Over the years the spotter format has been developed to include other subjects, to reduce the number of identify questions and increase the testing of other areas such as function and application.

Objective Structured Clinical Examination (OSCE)

OSCE is an assessment tool in which the components of clinical competence such as history taking, physical examination, simple procedures, patient management problems, communication, and attitude are tested using agreed checklists and rotating the students round a number of stations some of which have observers with check lists. OSCE is applicable to any situation where clinical competencies are to be tested. It is important to appreciate that there are major differences between the "spot test" and "OSPE". These differences are highlighted in Table 1. The clinical application of anatomical knowledge to patient care is evaluated mainly through OSCE. The main differences between OSPE and OSCE are also listed in Table 2.

Table 1. Differences between OSPE and Spot Examination

		OSPE	Spot Examination
1	Domain	Knowledge (analytical and interpretational)	Knowledge (Only recall)
2	Number of objective / station	Only one	No limit
3	Tests ability to	Identify structures on cadaver and radiographs and relate it with clinical scenario	Identify structures on cadaver, radiographs etc. only
4	Integration	Essential	Not possible
5	Variety of questions	Very high	limited
6	Blue printing	Essential	Not in accordance with blue print

Table 2. Differences between OSCE and OSPE

		OSPE	OSCE
1	Domain	Higher level of Knowledge	Psychomotor
2	Tests ability to	Identify structures on cadaver and radiographs and relate it with clinical scenario	Apply structural knowledge to perform examination and procedures
3	Time	2-4 minutes / station	>4 minutes / station
4	Space	One big hall	Isolated small rooms close to each other
5	Examiner	Nonspecific 2-3 personal per hall	Expert examiner for each station
6	Check List	Not required	Required
7	Standardized Patients (SPs)	Not required	Required
8	Observers	Not Required	Required

Viva-Voce or Oral Examination

It is an examination consisting of a dialogue with the examiner, who asks questions to which the candidate must reply. The oral examination is still used in some subject centered medical schools. Viva-voce or oral examination probe students more deeply than an essay, allows students to express more or less clearly their knowledge of isolated facts, assesses communication skills and professional attitude and if appropriately conducted, very useful for students as a formative assessment. Disadvantages of oral examination, on the other hand, are numerous; oral questions are not interrelated, questions and responses are not recorded, scoring practice is variable, fear of bias, students are expected to respond immediately to any question from a wide area of discipline, the environment is of confrontation rather than co-operation, most of the questions are of simple recall type, and interpretive and problem solving skills are difficult to test.

Objective Structured Practical Examination (OSPE)

The OSPE is a tool which is used to assess anatomy laboratory skills of students in the preclinical stage of a medical curriculum. OSPE is a new concept in practical assessment of basic medical sciences. It is a modified form of OSCE. OSPE is the main type of practical examination used in most of the PBL schools to assess knowledge and competence in anatomy instead of simple recall of memorized facts that occurs in the spot examination used in many traditional medical schools. In OSPE, multiple stations are designed and each station has a specific objective which needs to be tested. The objectives tested in an OSPE assess higher cognitive and psychomotor skills, i.e., ability to relate clinical information with structural material (plastic models, cadaveric specimen, photomicrographs etc.). The usual process of constructing an OSPE starts with identifying key objectives which need to be tested. A multidisciplinary approach is usually taken to write questions which tests core anatomical knowledge in functional and clinical context. Each station is set-up describing a clinical scenario and specific questions are formulated to test the objectives. Each station is equipped with appropriately labeled cadaveric specimens, models, images and physiological/pathological data sets to answer relevant questions. As with other practical exams, the major process which is involved is that the students must move from one station to the other in an orderly fashion. This approach, initiated approximately 30 years ago, has evolved and withstood the test of time and is now used by many medical schools to assess the practical skills of students in anatomy, physiology, pathology, forensic medicine etc. OSPE has been identified as the most useful method for assessment of anatomical knowledge at multiple levels of instruction: student, resident, specialist. OSPE is still considered as the most common and popular method of assessing practical or laboratory based basic sciences in the early phase of medical curriculum.

History of OSPE

OSPE is an extended form of OSCE which was described in 1975 and in greater detail in 1979 by Harden and his group from Dundee. OSPE was first introduced as a teaching and evaluation tool in 1986 by Nayar and colleagues to assess the practical skills of students in a physiology course.

Skills tested by OSPE

Besides testing high cognitive and psychomotor skills, OSPE helps to assess the capacity for observation, analysis and interpretation. OSPE can be structured to achieve the desired mix of different elements being assessed, each element receiving the desired weightage.

Why OSPE?

The OSPE is a popular method of assessing practical or laboratory based basic science in the early phase of medical curricula. First, OSPE revolves around a single objective; it allows the student to appreciate more on its purpose in a given scenario. Second, the newer schools can benefit more from its pedagogical values by placing minimum pass level ahead of time in a practical examination. Third, the OSPE allows to instituting different modalities available in 21st century including the traditional method of studying anatomy on cadavers, by dissection, manipulation, and exploration as well surface anatomy, medical imaging methods and therapeutic procedures such as open surgery, laparoscopy, and arthroscopy. Fourth, OSPE can certainly integrate such modalities.

Merits of OSPE

1. It is objective, reliable, valid and discriminatory
2. All students are exposed to same standardized questions
3. It covers a wide spectrum of learning domains
4. It tests a wide range of skills in a short period of time
5. Learning objectives can be achieved
6. The content and complexity of the exam can be controlled by the examiners
7. Gives a reasonable idea of the achievement of the student in every objective of practical exercises.
8. Helps to test the analytical abilities of the students
9. Organization is easy
10. Questions bank can be made
11. Large number of students can be examined in a short time.

De-Merits of OSPE

There are only a few disadvantages:

1. Substantial faculty hours are required to construct, review and develop integrated multidisciplinary questions,
2. The availability of quality cadaveric dissections, images, laboratory and physiological data sets.
3. It takes a lot of time, effort and team work on the part of examiners.
4. All stations must invariably demand equal time; therefore, it requires careful organization.

Organization of OSPE at AlFaisal (Yaqinuddin et al., 2012)

In PBL based medical curriculum, anatomy is taught in block system. In each block, contact time of anatomy resource sessions includes lectures, small group sessions (SGS) and laboratory sessions which are based on block objectives.

Overall, contact time for anatomy in the curriculum is approximately 200 hours. Approximately, 20-25% of the contact time is dedicated to lectures, 35-40% by the laboratory sessions and 35-40% by SGS. Anatomy is being taught in the laboratory sessions through: (a) cadaveric pro-sections, (b) plastic models, (c) radiological images, (d) microscopic slides and (e) dissection videos. OSPE at AlFaisal University consists of 20-24 functional stations (2-3 minutes each) for the examination of a particular block.

Table 3. Modified Ebel's method

	Relevancy/Difficulty	Raw Score (A)	Proportion of borderline students who can answer correctly (B)	Expected Score= A x B x 100
Station-1 Judge -1	Essential Easy	1*	100	
	Essential Medium	0.9	90	
	Essential Difficult	0.8	80	
	Important Easy	0.7	70*	
	Important Medium	0.6	60	
	Important Difficult	0.5	50	
	Marginal Easy	0.4	40	
	Marginal Medium	0.3	30	
	Marginal difficult	0.2	20	
			10	
	Total			70%

In a Modified Ebel's method several judges are asked to fill a matrix for each question of the OSPE based on relevance or difficulty and proportion of borderline students who can possibly answer the question correctly. In this matrix, Raw score 'A' is the score for relevance-difficulty, 'B' is the proportion of borderline students and the last column of the matrix displays expected score (A x B).

The questions for each station are written in accordance with the blue-print of the block. Organization of OSPE for the "end of block" examination is such that there may not be a repetition of objectives tested in OSPE, MCQ and SAQ portions.

The answer key is prepared before examination. The OSPE questions (stations) are reviewed by a multi-disciplinary committee before they are administered. Modified Ebel's method is in use to identify the minimum pass level (MPL) of the OSPE.

To calculate MPL, three to six judges (anatomist with some clinical background) are selected. These judges are asked to categorize the OSPE questions according to the relevance and difficulty. Secondly, they are asked to discuss and estimate the proportion of borderline students that will be able to answer the questions correctly.

An expected score at each question is calculated by multiplying raw score for relevance, or difficulty is multiplied with proportion of minimally competent students in order to calculate an expected score. Average of the expected scores from the judges is than taken to calculate the MPL for each station, an example is provided in Table-3. A clinical scenario is given at each station and students are asked integrated questions. These would incorporate knowledge components including structural, functional, clinical, radiological and microscopic aspects. For example, for a given clinical scenario students are asked to: (a) identify the labeled or tagged structures involved, (b) interpret physiological/pathological laboratory data and (c) identify the labeled structure on a radiograph.

Blue print for OSPE

While preparing a blue print or table of specification for the OSPE of a particular course, due weightage should be given to each objective, and due representation should be given to every aspect of the course. A faculty member is assigned as the resource person for the particular unit/block. His/ Her responsibility is to write questions in accordance to specific objectives to be tested.

These objectives should be carefully prepared by the Unit/block committee through an OSPE blue print. Generally, the blue print of the OSPE is developed as a part of the master blue print, which comprehensively describes the objectives which need to be tested in the end of block examination. It explains the method of testing e.g. MCQ, SAQ, OSPE and the taxonomy level. Thus a block committee adopts the following sequence: (1) identifies the important learning needs (2) formulates a blue print and (3) calls for a meeting with the appropriate resource persons including anatomist, physiologist and clinicians.

Based on a given blue print the resource person in-turn presents OSPE questions to the block committee. The members through a concerted effort correct the questions as needed and also identify modalities (cadaver, plastic model, drawing, histology, CT/MRI/X-ray, plastinated model, living subjects) to be used for these questions. The overall deliberations ensure clarity, sound factual content and appropriate emphasis on the clinical application.

This exercise ensures that the questions are not ambiguous and the students are able to answer each question within time limit. A typical OSPE blue print for a musculoskeletal block is given (Table 4). A typical layout of the musculoskeletal block OSPE with logistic requirements is given (Figure 3). Generally, most anatomy stations are not only integrated with physiology but also test skills of identification, interpretation of results in clinical context.

Finally, the questions are sent to the Phase Director who appoints three to six judges who evaluate the questions and set up minimum pass level using Ebel's method as described earlier. The questions are then handed over to assessment office for execution.

Table 4. Blue-print of OSPE of the Musculo-skeletal Block

	Cadavers	Plastinated Models	Plastic Models	Microscopic slides	Photomicrographs	CT/MRI	X-Ray	Embryonic models	Total
Identification of tissue / organ				3					3
Identification of tissue / organ with functional correlation					3				3
Development of limbs with anomalies								2	2
Identification of muscle (actions with clinical applications)	4	1	1			1			7
Identification of nerves with clinical applications	2	1	1						4
Identification of bony landmarks / fractures							1		1
Joints						1	1		2
Miscellaneous	1		1						2
									24

The number represents OSPE station.
Each OSPE station is carefully developed taking into account functional or clinical anatomy.

To evaluate the assessment done via OSPE, item analysis for each station/question is performed. This includes calculating mean score at each station. At any given station/question a mean score of less than 50% is reviewed by the multidisciplinary OSPE organizing committee. The questions which are found to be ambiguous are dropped.

Each OSPE station usually has two parts: (A) related to identification of a structure and (B) a secondary question involving application of knowledge. For example in Figure 1 a clinical scenario of a 55 year old male with squamous cell carcinoma and thickening of skin is given.

Accordingly, the OSPE station shows a photomicrograph pointing “D” as stratum basale, “C” as stratum spinosum, “B” as stratum granulosum, “A” as corneum and “E” as dermis. The first part of the question requires identification of the layer “B” in which the cells are squamous in shape and give origin to squamous cell carcinoma of skin. The second part of the question requires the knowledge of the keratinization process.

Students are expected to know the normal transit time of the cells from their origin in basal layer to desquamation from stratum corneum. This is important to understand a disease called Psoriasis in which this time is significantly shortened to about 7 days. Thus, the students not only identify structures but also correlate their importance functionally and clinically. We are also in the process of developing "observed stations" in anatomy to test psychomotor skills in OSPE. These observed stations with living subjects are very helpful, particularly in musculoskeletal system to test skills such as examination of joint movements or in nervous system for function of cranial nerves.

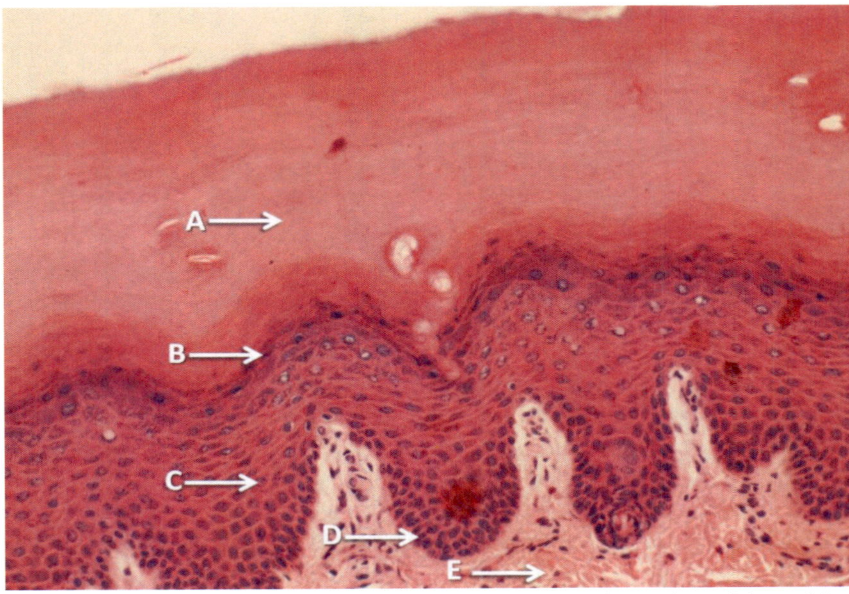

Questions: 1) Which of the following lettered cell layers is more likely to give origin to this carcinoma?
2) **What is the normal transit time for the cells in layer "D" through "A"?**

Key:

A: Squamous cell carcinoma originates from stratum granulosum labelled as B.

B. The transition of keratinocytes from replicating basal cells through the keratinized layer takes 50-60 days.

Figure 1. A 55-year-old male was suffering from squamous cell carcinoma and thickening of skin.

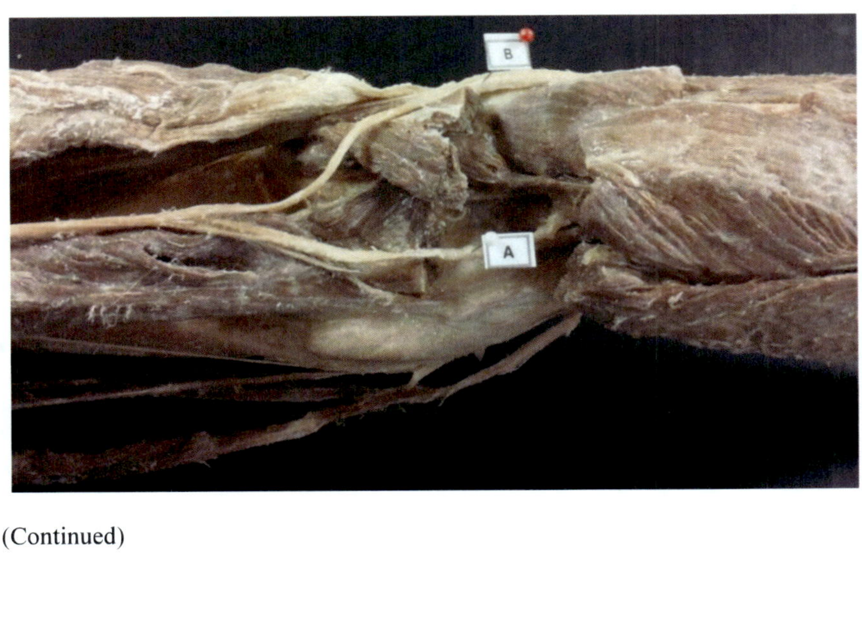

Figure 2. (Continued)

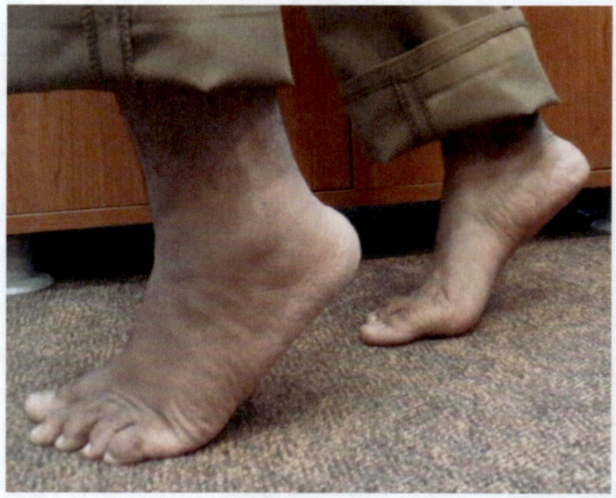

Question: 1) Which of the marked structure A or B on plastinated lower limb is responsible for this movement? 2) Which of the marked structure A or B on plastinated lower limb is liable to be injured in case of fracture of neck of lateral leg bone?.

KEY:

1. Answer is A which is Tibial nerve which supplies the long flexor muscles of leg which cause planter flexion.
2. Answer is B which is common peroneal (fibular) nerve which winds round the neck of fibula hence liable to be injured

Figure 2. Plastinated Lower Limb. A physician asked his patient to stand on the tip of his toes as shown in the picture.

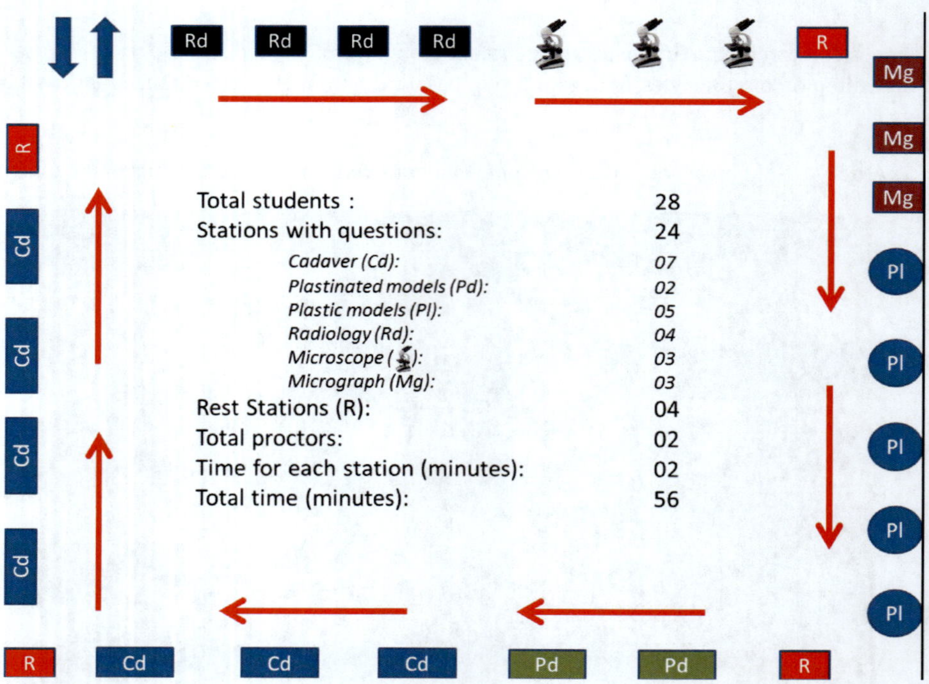

Figure 3. Illustrates a layout of a typical OSPE at AlFaisal University.

Precautions for the Organization of OSPE

1. Preparation for OSPE should start 6-8 weeks before the date of examination
2. There should be a coordinated activity of staff at all levels
3. Objectives tested with MCQs and SAQs should not be repeated

Instructions to Students

Before the start of OSPE, following instructions are given to students:

1. Bring your own pen/pencils and must wear white lab coats
2. Electronic devices like mobile phones, tabs etc. are not allowed in the examination hall
3. Cheating in any respect is strictly prohibited and the regulation of the university will be applied
4. Write your name and ID on each sheet of answer books
5. Upon entering the OSPE examination hall, stand on each station with face opposite to station
6. Encircle the station number in your answer books. This will be your first station and then follow the sequence
7. A bell will ring at the beginning of OSPE marking the start of examination
8. Rotate clockwise around 28 stations including 4 rest stations spending 90 seconds (1.5 minutes) at each station
9. Clear instructions are given at each stations as to what you should do
10. Do not touch or move the models, specimens or plastinated specimen
11. Do not move the microscopic slides, only fine adjustments knobs can be used
12. A bell will ring at the end of examination and remain at your station from where the answer books will collected

Is OSPE a Valid and Reliable Tool?

OSPE is more objective, reliable and a valid tool of assessment to assess practical aspects of anatomy in an integrated problem based curriculum. It is designed in such a way as to get a reasonable idea of the achievement of student in every objective of practical exercises. Validity refers to the ability of the instrument to measure what it is supposed to measure. A valid instrument accurately reflects the extent to which the student has achieved the objectives of the course. When the number of test items is more, the assessment is more valid. OSPE fulfills this criterion. In order to maintain the validity of OSPE, the time for each station should be same (2-4 minutes per station). Further, in a particular OSPE of a single course, all stations should be completed in the same period of time. The students must rotate through all stations and have to move to the next station at the bell.

Reliability refers to the consistency of the assessment tool. An objective assessment is devoid of variability due to different examiners. Recent findings have shown that the scores in multiple choice tests and short answer questions correlate strongly with OSPE scores and a highly integrated OSPE is an objective and effective tool to assess anatomical knowledge in a PBL curriculum. In OSPE the process as well as the product is tested giving importance to individual competency. Validity and reliability of OSPE is improved by preventing examiner variability. Validity and reliability of OSPE is also enhanced by the employment of Ebel's method, blue printing of the learning outcomes, and evaluation and statistical analysis of a range of elements such as individual questions in an examination.

CONCLUSION

OSPE is a new concept in basic medical sciences. Over the past twenty years, the major method of assessment in anatomy has been written examinations including MCQs, EMQs and SAQs. However, there are strong recommendations to assess anatomy through practical methods. In this regard, many methods have been adopted including cadaveric spot tests; computer based spot tests, web-based laboratory exercises, oral examinations and OSPEs. A highly integrated OSPE provides an opportunity to assess clinical application of anatomical knowledge. It has been shown previously that use of such type of assessment tool promotes an integrated approach to learning anatomy and increases student understanding. OSPE tests different desired components of competence better. It is an objective, valid and reliable method and gets rid of variation due to examiner bias. It has a better discrimination index on merit. Students feel more comfortable and less stressed to perform the exam.

Curriculum at AlFaisal addresses integration of structure and function of human body at the preclinical phase. Students discuss this objective and related anatomy content through weekly PBL cases within small-group sessions. The major Anatomy courses are taught in first four semesters (phase 1). The subject of Anatomy is revisited at the early part of each block in Phase 2 (two semesters) mostly to understand abnormal structures. Phase 3 of our curriculum (final four semesters) is directed towards clinical subjects and anatomy plays a role to conceptualize diagnosis of a disease process. However, at any point of our curriculum, significant efforts have been made to integrate anatomy with other basic and clinical sciences. Therefore, laboratory sessions are geared towards objectives that not only can relate to osteology, imaging, cadaveric dissection, histology, embryology, neuro-anatomy and surface anatomy, but also to integrate pathophysiology and clinical aspects of the disease process.

In spite of its limited drawbacks, OSPE is a great tool once the curricular objectives are primarily related to integration with clinical anatomy. Therefore, incorporating OSPE into the medical curriculum is ideal in a PBL setting. In principle, our OSPE appropriately tests students' grasp of anatomical knowledge, and we believe it is motivating, inspiring and interesting. OSPE should follow the curriculum and demands higher cognitive skills. Therefore, there is a need to develop objective tools to assess higher cognitive skills like clinical application of anatomical knowledge.

REFERENCES

Abu-Hijleh M, Chakravarty N, Latif M, Osman A, Salem R, Fadel and A. Dharap. 2009. *The place of Anatomy in Medical Education*, Amee guide supplement 41.2.

Agarwal A, Batra B, Sood AK, Ramakantan R, Bhargava SK, Chidambaranathan N, Indrajit IK. 2010. Objective structure clinical examination in radiology. *Indian J. Radiol. Imaging* 20:83–88.

Ananthakrishnan N. 1993. Objective structured clinical/practical examination (OSCE/OSPE). *J. Postgrad. Med.* 39:82–84.

Aarti S M, Nilima S, Tandon O P.2012. The Comparison of OSPE with conventional Physiology Practical Assessment. *JIAMSE* vol.14:No 2-Articles.

Azeem MA. 2007. A brief overview regarding various aspects of objective structured practical examination (OSPE): Modifications as per local needs. *Pak. J. Physiol.* 3:1–3.

Barman A. 2005. Critiques on the objective structured clinical examination. *Ann. Acad. Med. Singapore* 34:478–482.

Bowe CM, Voss J, Aretz TH. 2009. Case method teaching: An effective approach to integrate the basic and clinical sciences in the preclinical medical curriculum. *Med. Teach.* 31:834–841.

Chakravarty M, Latif NA, Abu-Hijleh, MF, Osman M, Dharap, AS, Ganguly PK. 2005. Assessment of anatomy in a problem-based medical curriculum. *Clin. Anat.* 18:131–136.

Chan LK, Ganguly PK. 2008. Evaluation of small group teaching in human gross anatomy in a Caribbean medical school. *Anat. Sci. Edu.* 1:19–22.

Collins TJ, Given RL, Hulsebosch CE, Miller BT. 1994. Status of gross anatomy in the U.S. and Canada: Dilemma of the 21st Century. *Clin. Anat.* 7:275–296.

Cowan M, Arain NN, Assale, TS, Assi AH, Albar RA, Ganguly PK. 2010. Student-centered integrated anatomy resource sessions at Alfaisal University. *Anat. Sci. Educ.* 3:272–275.

Dissanayake AS, Ali BA, Nayar U. 1990. The influence of the introduction of objective structured practical examinations in physiology on student performance at King Faisal University Medical School. *Med. Teach* 12:297–304.

Downing SM, Tekian A, Yudkowsky R. 2006. Procedures for establishing defensible absolute passing scores on performance examinations in health professions education. *Teach Learn Med.* 18:50–57.

Drake RL, McBride JM, Lachman N, Pawlina W. 2009. Medical education in the anatomical sciences: The winds of change continue to blow. *Anat. Sci. Educ.* 2:253–259.

Eldred E, Eldred B. 1961. Supply and demand for faculty in anatomy. *J. Med. Educ.* 36:134–147.

Feroze M, Jacob AJ. 2002. OSPE in pathology. *Indian J. Pathol. Microbiol.* 45:53–58.

Harden RM. 1988. What is an OSCE? *Med. Teach* 10:19–22.

Harden R M, Stevensen M, Downie W W, Assessment of clinical competence using objective structured examination. *Br. Med. J.* 1975 Feb;1(5955):447-51

Harden R M, Gleeson F A. Assessment of clinical competence using an objective structured clinical examination (OSCE). *Med. Educ.* 1979 Jan;13(1):41-45.

Hartling L, Spooner C, Tjosvold L, Oswald A. 2010. Problem-based learning in pre-clinical medical education: 22 years of outcome research. *Med. Teach* 32:28–35.

Hasan S, Malik S, Hamad A, Khan H, Bilal M. 2009. Conventional/Traditional Practical Examination (CPE/TDPE) versus Objective Structured Practical Evaluation (OSPE)/Semi Objective Structured Practical Evaluation (SOSPE). *Pak. J. Physiol.*;5(1)

Hofer RE, Nikolaus OR, Pawlina W. 2011. Using checklists in gross anatomy laboratory improves learning outcomes and objectives. *Anat. Sci. Educ.* 4:249–255.

Inuwa IM, Taranikanti V, Al-Rawahy M, Habbal O. 2011. Perceptions and attitudes of medical students towards two methods of assessing practical anatomy knowledge. Sultan Qaboos Univ. *Med. J.* 11:383–390.

Inuwa IM, Taranikanti V, Al-Rawahy M, Habbal O. 2012. Anatomy practical examinations: How does student performance on computerized evaluation compare with the traditional format? *Anat. Sci. Educ.* 5:27–32.

Latif SA, Hossain M. 2004. Objective structured practical examination (OSPE) - a scheduled task in curriculum for undergraduate medical education in Bangladesh - 2002. *Mymensingh Med. J.* 13:100–101.

Levine MG, Stempak J, Conyers G, Walters JA. 1999. Implementing and integrating computer-based activities into a problem-based gross anatomy curriculum. *Clin. Anat.* 12:191–198.

Lukic IK, Gluncic V, Katavic V, Petanjek Z, Jalsovec D, Marusic A. 2001. Weekly quizzes in extended-matching format as a means of monitoring students' progress in gross anatomy. *Ann. Anat.* 183:575–579.

Menezes RG, Nayak VC, Binu VS, Kanchan T, Jagadish Rao PP, Baral P, Lobo SW. 2011. Objective structured practical examination (OSPE) in forensic medicine: Students' point of view. *J. Forensic Leg. Med.* 18:347–349.

Munjal K, Prasana K B, Amit V and Shirish N. 2012 January. Assessment of medical students by OSPE method in pathology. *Internet Journal of Medical Update*;7(1):2-7.

Nayar U, Malik SL, Bijlani RL. 1986. Objective structured practical examination: A new concept in assessment of laboratory exercises in preclinical sciences. *Med Educ.* 20:204–209.

Pawlina W. 2009. Basic sciences in medical education: Why? How? When? Where? *Med. Teach* 31:787–789.

Philip CT, Unruh KP, Lachman N, Pawlina W. 2008. An explorative learning approach to teaching clinical anatomy using student generated content. *Anat. Sci. Educ.* 1:106–110.

Rahman N, Ferdousi S, Hoq N, Amin R, Kabir J. 2007. Evaluation of objective structured practical examination and traditional practical examination. *Mymensingh Med. J.* 16: 7–11.

Rowland S, Ahmed K, Davies DC, Ashrafian H, Patel V, Darzi A, Paraskeva PA, Athanasiou T. 2011. Assessment of anatomical knowledge for clinical practice: Perceptions of clinicians and students. *Surg. Radiol. Anat.* 33:263–269.

Schmidt H. 1998. Integrating the teaching of basic sciences, clinical sciences, and biopsychosocial issues. *Acad. Med.* 73:S24–S31.

Schoeman S, Chandratilake M. 2012a. The anatomy competence score–a new marker for anatomical ability. *Anat. Sci. Educ.* 5:33–40.

Schoeman S, Chandratilake M. 2012b. The weak relationship between anatomy competence and clinical skills in junior medical students. *Anat. Sci. Educ.* 5:000–000 (in press; doi: 10.1002/ase.1272).

Torke S, Upadhya S, Abraham RR, Ramnarayan K. 2006. Computer-assisted objective-structured practical examination: An innovative method of evaluation. *Adv. Physiol. Educ.* 30:48–49.

Vasan NS, DeFouw DO, Compton S. 2009. A survey of student perceptions of team-based learning in anatomy curriculum: Favorable views unrelated to grades. *Anat. Sci. Educ.* 2:150–155.

Vasan NS, DeFouw DO, Compton S. 2011. Team-based learning in anatomy: An efficient, effective, and economical strategy. *Anat. Sci. Educ.* 4:333–339.

Vasan NS, DeFouw DO, Holland BK. 2008. Modified use of team-based learning for effective delivery of medical gross anatomy and embryology. *Anat. Sci. Educ.* 1:3–9.

Yaqinuddin A, Kvietys P, Ganguly P, Ikram F, Yaeesh S, Kattan W. 2012. PBL performance correlates with content acquisition assessment: A study in a hybrid PBL program at Alfaisal University. *Med. Teach* 34:83.

Yaqinuddin A, Zafar M, Ikram MF, Ganguly P. What is an objective structured practical examination in anatomy? 2012 Aug 28. doi: 10.1002/ase.1305. [Epub ahead of print]

Yiou R, Goodenough D. 2006. Applying problem-based learning to the teaching of anatomy: The example of Harvard Medical School. *Surg. Radiol. Anat.* 28:189–194.

NOTES ON CONTRIBUTORS

Muhammed Zafar, M.B.B.S., M.Phil., Ph.D., is an associate professor of anatomy at the College of Medicine, AlFaisal University, Riyadh, Kingdom of Saudi Arabia. He teaches anatomy to medical students in the first, second and third semester. His research interest is in curriculum development.

Ahmed Yaqinuddin, M.B.B.S., Ph.D., is an assistant professor of anatomy at the College of Medicine, AlFaisal University, Riyadh, Kingdom of Saudi Arabia. He teaches anatomy to medical students in the first, second and third semester. His research interest is in curriculum development.

Muhammed Faisal Ikram, M.B.B.S., M.Phil. is a lecturer of anatomy at the College of Medicine, AlFaisal University, Riyadh, Kingdom of Saudi Arabia. He teaches anatomy to medical students in the first, second and third semester. His research interest is in assessment and applying new technologies in teaching.

Paul Ganguly, M.B.B.S., M.D., F.A.C.A., is a professor and chairman of the Department of Anatomy at the College of Medicine, AlFaisal University, Riyadh, Kingdom of Saudi Arabia. He teaches anatomy to medical students in the first, second and third semester. His research interest is in curriculum development and assessment.

In: Education in Anatomical Sciences
Editor: Paul Ganguly
ISBN: 978-1-62618-488-6

Chapter 17

EDUCATION IN ANATOMY: STUDENTS' PERCEPTION

Abdulhadi A. AlAmodi, Ahmed Abu-Zaid, Akram M. Nurhussen, Ahmed A. Shamia, Tehreem A. Khan, Riya Ganguly, Matthew S. Lytwyn and Paul Ganguly*

College of Medicine, Alfaisal University, Riyadh,
Kingdom of Saudi Arabia and St. Boniface Research Centre,
Winnipeg, Canada

ABSTRACT

Anatomy is integral to the education of medical students. Teaching and learning anatomy has been a subject of debate and controversy among traditionalists and modernists especially in the light of a worldwide shift in adopting a student-centered approach. There is emphasis on embracing newer modalities in teaching and learning anatomy, and engaging medical students in decisions related to curriculum development and plan. However, arguments emerge warning that the knowledge of anatomy might decline. Students' perceptions and inputs are of a great value since this issue primarily concerns their education and development. In this chapter, we present a thorough literature review one ducation and anatomy from a student's perspective. In this context, we elaborate on three areas; anatomy curriculum, methods employed in teaching anatomy and anatomy assessment. We also present Alfaisal University as an illustration of anatomy in medical education using a structured questionnaire and previously published literature to prove our findings.

GLOBAL SHIFT IN THE PARADIGM OF MEDICAL EDUCATION DELIVERY

Medical knowledge, technology and expansion of specialization have mandated exploration of new approaches to overcome evolving challenges in medical education. The

* E-mail Address: pganguly@alfaisal.edu.

traditional approach in designing medical curricula and teaching medical students is gradually fading worldwide. There is an increasing global shift into the model of integrating basic medical sciences [1, 2]. Recognized progress in adopting integrated medical curricula has also been achieved in many countries including Korea, Japan, Sweden and Italy [3-6].

Considerable application of the SPICES model (Student-Centered, Problem-Based, Integrated, Community-Based, Elective and systematic) in the Gulf Cooperation Council has been achieved in recent years: since inception, most of the newly established medical schools in the Kingdom of Saudi Arabia and other Gulf countries have structured their curricula based on the SPICES model. Moreover, medical schools currently adopting the traditional model have also shown interest towards integrating the SPICES model [7-9].

Students As Agents of Change in Medical Education

Curriculum design and development is a continuous process that requires input from all concerned parties including, and perhaps primarily, faculty members. The role of students in curriculum development has recently been questioned [10-13]. There are two views regarding the involvement of students in the decision making process of curriculum design and development. The first view does not support students' participation in curriculum planning. This view is mainly adopted by traditionalists who argue that students' engagement in decision making to curriculum plan would result in distracting the process of learning, and encourages students to suspend teaching for their own interests and purposes [11]. Conversely, the literature suggests that the opposite view has recently mounted. The argument encourages students' involvement in curriculum development since the purpose of curricula is to promote their growth, development, and education. Studies have demonstrated that students favor an active educational environment that supports problem solving skills [10-15]. This argument becomes intensified with the in-progress initiatives geared towards reforming medical curricula. D'eon and Harris have argued in a comprehensive article entitled "If students are not customers, what are they?" that students play a fundamental role in making decisions related to issues that greatly influence their learning [13-15]. Alfaisal University has shown to support active student roles in all matters and issues pertaining to their education and learning. Students are constantly encouraged to elect and appoint representatives who possess dynamic roles by presenting issues, suggestions, comments and feedback in different committees' meetings such as: Curriculum Committee meeting, Assessment Committee meeting, and College Board meeting. Another adapted approach is to obtain students' inputs via constructed instrumental tools and surveys regarding course evaluation, assessment evaluation and faculty evaluation.

Student's Perception of Anatomy in Conventional and Integrated Medical Curricula

Despite a global shift in medical education delivery to the integrated medical curricula, the effectiveness of integrated medical curriculum over conventional medical curriculum remains debatable and needs further investigation [16, 17]. The advocates of conventional medical curriculum argue that students' outcomes and scores from standardized assessments are not significantly different from students from the integrated medical curriculum [18]. There is also a perception that a student-centered approach might lead to the production of medical students with insufficient anatomy knowledge [19-21]. In contrast, effective practice of medicine requires medical students to develop skills and attitudes beyond mere standardized assessment [22]. Anemerging and well-studied method in modern medical curricula is the Problem-based Learning module (PBL). Students exposed to medical curricula supporting Problem-Based Learning modules have shown to exhibit better competency in history-taking skills, working in groups, solving problems, becoming self-directed learners and possessing long term recall of information [23-26]. Meta-analysis research studies have shown that PBL is a superior approach over more traditional methods [27]. Yet, the application of PBL to teaching and learning anatomy has been a subject of controversy [16]. This might be attributed to the nature of anatomy as a subject to be defined overtly in terms of what to be taught, when to be taught and how to be taught [19].

While a number of studies have examined medical students' perception regarding learning anatomy in traditional and modern medical curricula, an important factor which medical educators and instructors need to consider is the students' prior knowledge about the subject of anatomy. Students are classified into three categories in terms of learning which include deep approach, surface approach and strategic approach. Students who tend to follow the deep approach to learning construct a strong base of understanding and long term recall of information. It is achieved by utilizing high order cognitive skills which include integrating ideas and concepts together, thinking and analyzing critically, while those who subscribe to the surface approach learning theory exhibit superficial retention of knowledge and short term recall of information due to memorizing and acceptance of facts without any effort to link or integrate them [28-30]. One study examined students' perception and experiences in learning anatomy and its association with the students' approach to learning the subject. Students who exhibit the deep approach to learning effectively understand the subject through dissection labs, understanding the clinical relevance of anatomy and correlating anatomical structures on different radiology modalities. On the other hand, students who prefer the surface approach have perceived anatomy as a dull subject and not beneficial. This study indicates that moving into an integrated student-centered approach will be effective by promoting students to apply the deep approach to learning [30]. Another study examined first-year medical students' perception of learning gross anatomy through a traditional approach vs. the problem-based (PBL) approach. Consequent results showed that first-year medical students perceived the problem-based learning approach as a more effective means of learning anatomy; students have reported that they gained more anatomy knowledge through PBL in comparison to the traditional approach with one of the key drawbacks of the latter cited as the lack of interactive group discussions [31].

In conclusion, the majority of emerging studies which have examined students' perceptions regarding learning patterns in anatomy, show that students largely prefer active learning and integrated curricula over the traditional approach. However, further studies are required.

STUDENTS' PERCEPTION ABOUT METHODS EMPLOYED IN TEACHING AND LEARNING ANATOMY

Anatomy in the Current Century

Current anatomy teaching modalities range from the use of atlases, books, dissections, prosections, and plastinated specimens to the use of radiological modalities, computer assisted learning and internet web pages [32–35]. The invention of new technologies has added a spectrum of new techniques used to teach anatomy in the modern era. As a result, medical colleges have a wide range of options for teaching devices for students. All of these modalities are capable of delivering information; however there are no studies to date ranking the efficacy of one modality over the other [36].

Students' Perception about Dissection

There is a significant discrepancy between the use of the aforementioned modalities and their use by anatomists' [19]. For example, European anatomists' view anatomy learning is best achieved by dissection [37] while students in Australian schools believe that both dissection and textbook learning are the best modalities in achieving knowledge of anatomy [38]. Similarly, students of British schools also view dissection as the best fit method for teaching anatomy [32]. Conversely, a study conducted in Berlin assessed the cognitive impact of dissection on learning outcomes and revealed that all studies which took part in their review lacked consensus. Although dissection had an advantage, the author found it difficult to ascertain whether dissection is the best fit for teaching anatomy and advised organized researches in order "to come to scientifically sound conclusions about the best way to teach gross anatomy" [39].

Briefly, the term "dissection" is defined as "the act or process of dissecting or separating" [40]. Dissection not only aids in gaining knowledge but also develops hands on experience, a sense of cooperation, group discussion, "respect for the human body", and professionalism amongst students [41]. The cadaver used in the dissection lab is also considered as the students' "first patient" [42]. Many medical schools have found methods which make this modality as beneficial as possible. For example, the University of Otago in Dunedin, New Zealand, organized a dissection competition where the participants were third-year medical students. The competition yielded educational knowledge among participants in addition to the production of well-prepared prosection samples to be used for anatomy classes [43].

Dissection laboratories may take several hours to be completed which can be a dreary task [41].A Korean study wanted to make the dissection session exciting and support discussion among students. The study organized sessions in which students were given a set of tasks to be answered solely by the aid of the dissection. The study concluded that the applied method did not enhance knowledge of anatomy however team discussion and cooperative learning among students improved [44].

Dissection and the dissection lab environments are stressful especially for first-year medical students. The presence of a lifeless body, irritating odors emerging from the cadaver and the experience of dissecting potentiates a great deal of stress [45-47]. The most commonly observed reaction to such environment is "headache, disgust, grief or sadness, and feeling light-headed" [46]. Anxiety is also not uncommon, therefore student preparation for dissections is necessary [47]. Since all students do not possess similar personality traits, it is important to assess their personalities [48] and provide support, in order to facilitate getting the most of the learning experience during the dissection session [47].

Time in the laboratories is also an important factor to consider. Less time in dissection was among the factors which were found to yield less understanding of anatomy and therefore negatively affect outcome on standardized written tests as compared to students who practiced dissection regularly [49]. One approach to benefit from dissection sessions is to have a known set of objectives prior to each dissection. In this way students become more familiar regarding requirements of a specific session in terms of acquiring knowledge. A study conducted by Mayo Clinic focused on this issue using a daily check-list of objectives. This study found better understanding and retention of anatomy, compared to when check-lists were not implemented [50].

Team Based Learning in Teaching Anatomy

Small group discussions and team work is a key component of Team-Based Learning (TBL) along with "out of class" material study by the group [51]. All these methods make TBL an exceptional way of learning with greater emphasis and responsibility on the student rather than the faculty. TBL depend on 4 major aspects to function properly and these are: "Groups, Accountability, Feedback and assignment design" [52]. TBL has a great role in teaching anatomy as well. The time spent in lecturing is replaced with TBL sessions where students get exposed to the assignments. A study conducted in New Jersey Medical School, noted that students with a TBL system had a better outcome compared to the traditional system. They performed better in their internal exams and on the exams offered by NBME [51]. Similarly, a study conducted in Australia also showed a similar finding. Senior students were given an anatomy course using TBL and it was noted that the students' "post course assessment" had a major improvement compared to the pre-course. The knowledge of anatomy improved and the students agreed that TBL and its pillars were the reason for this improvement [53]. TBL had positive rating as a teaching modality by the students irrelevant to their academic score [54]. The students' grades were not the only thing that improved. The application of TBL also led to the development of "self directed learning", "peer respect", and being a "responsible team member "among the students [55].

Radiology as Means of Teaching Anatomy

The employment of radiological imaging and involvement of radiologists is well known for teaching anatomy in medical schools [34]. Radiology provides an excellent mode for viewing the spatial relationship of anatomical structures [56]. However, the extent to which radiology is utilized for teaching anatomy varies significantly [34]. Radiology is largely unexplored in teaching preclinical medical students [57]. Only 5 percent of anatomy teaching is achieved through the use of radiology [58]. Thus, the limited role of radiology in anatomy teaching is evident.

Furthermore, anatomy is usually offered in the preclinical years whereas radiology and surgery are offered as clinical years' subjects. It was postulated that since there is a significant time gap between these years, the knowledge of anatomy could get affected. The integration of radiology and surgery for teaching anatomy for 2nd year students yielded encouraging results as noted by a German study. It aided the students' understanding of anatomy. In addition, it gave the students a flavor of radiology and surgery which raised the students' interest in the subjects [59].

It is interesting to note that a German study utilized an unconventional method to teach anatomy by use of radiological images for both preclinical and clinical students. The special aspect in using this method was that the actual radiological images were processed by computer software resulting in three dimensional images. The students rated this method as beneficial for understanding anatomy and the topographic organization of the structures. In addition, there was greater consensus that it should be offered in the first year and recommendation for its continued implementation [60].

A study was conducted by Boston University School of Medicine to test the effectiveness of using cadaveric CT scans on the outcome of medical students' performance. It found that students who utilized the CT images performed better on standardized examinations compared to those who did not; concluding the importance of using such a technique in ascertaining positive outcome [56].

One of the major aims of teaching anatomy is to yield knowledgeable doctors who can apply medical knowledge to clinical practice. Knowledge acquired by students during undergraduate years can be used in different aspects including reading radiological images. The incorporation of radiology in teaching anatomy could have dual benefits for the students; in understanding anatomy as well as preparing them for reading, interpreting and understanding imaging in their future clinical practice [61]. Although radiology cannot substitute the traditional teaching of anatomy achieved by the use of cadaveric dissections, it could help complement more traditional methods, enabling medical students to gain sound knowledge of the subject [62].

Does Surgery Play a Role in Teaching Anatomy?

Although surgeons possess experience in teaching residents, they are not frequently utilized for teaching undergraduate students [63]. This dearth can be addressed as surgeons could help to assist the anatomist, and provide anatomy knowledge to the students within a clinical context [64]. Similarly, students need to be taught anatomy oriented to the more clinical aspects of medicine. In light of these beliefs and perceptions, a program was designed to incorporate a clinical surgery program to an anatomy course where one of the objectives was to know whether students would have better academic results. The students enrolled in the program had "the highest average overall for all combined anatomy examinations" [65].

Likewise, students taught by surgeons and participated in a surgical procedure reported this anatomy teaching experience as beneficial. This methodology has aided to increase the comprehension of anatomy and the clinical correlation of this subject [66].

In addition to this, not only would the students get a deeper understanding of anatomy, but also surgery would be considered as a future career choice. Majority of students agreed that this surgical approach in teaching anatomy, has aided them in "understanding of the knowledge, skills, and qualities needed to become a surgeon" [67]. Students had a positive view on surgeons and students' interest for surgery as a career was also enhanced. [67]

The use of laparoscopic surgery in teaching gross anatomy to first-year students is receiving positive reviews from them [68, 69]. Laparoscopic surgeries were performed on fresh cadavers by surgeons [70] and the students aided in the procedure [69, 70]. As a consequence, students received the chance to understand gross anatomy by using and maneuvering laparoscopic instruments [70]. The video recording of the procedure was displayed on monitors in an attempt to allow all the students to participate [69]. In the aftermath of their experience, students agreed that the use of laparoscopic surgery was greatly beneficial in understanding gross anatomy [70]. This approach allowed the students to not only appreciate the information they gained during their anatomy course, but it also served as helpful adjunct for teaching anatomy [69].Anatomy taught by surgeons could have multiple benefits. The students learning experience [65, 66] and performance in anatomy would increase [65], and they would be more oriented to surgery as a career choice [67, 70]. Students will also get the chance to work with real surgical instruments [70].

TECHNOLOGY AND OTHER MISCELLANEOUS TOOLS FOR TEACHING ANATOMY

Recent rapid advancements in technology have led to the production of a technology-oriented generation. Medical students of this generation have sound understanding and knowledge of technology, using it for a large variety of tasks, academic and otherwise [71]. Anatomy teaching should persevere with this growth and develop alongside it. A US study employed the use of technology in a rather impressive manner by compiling radiology imaging in an online, easily accessible, user friendly electronic web system. This method was used in accordance with lectures and dissection's objectives and the students reported it to be a helpful adjunct in understanding anatomy [71].

Another unconventional yet unutilized tool for teaching anatomy is the use of video sharing websites such as YouTube. A study published in the Journal of Surgical and Radiological Anatomy concluded that YouTube is still deficient as means of learning anatomy and urges medical schools to upload more useful media to it [66]. In the same year, another study published in the Journal of Anatomical Sciences Education had an opposing review [72]. This study believed that YouTube has the potential to become a useful source of teaching since students resort to it as an information source. However it also urged the faculty to have YouTube channels so they can upload their material as well [72]. Conclusively, the use of technology for delivering anatomical knowledge still could be further utilized to reach its potential.

Plastinated specimens are growing means of teaching anatomy adopted by medical schools [73, 74]. Plastination is a process of preserving dissected specimens protecting them from decay. They are odorless, safely exposed in the lab, handled with bare hands and easily stored without the need of any special technique. Students perceive the use of plastinated specimen for learning anatomy as helpful [74]. According to students, plastinated specimen offers a wide range of advantages such as "the detailed view of relevant anatomy, appreciation of relations between structures, and visualization of anatomy in real life" [73]. Although useful, and valuable, the teaching of anatomy through the use of plastinated specimen could further be augmented by other teaching modalities such as dissection [73, 74].

Computer Assisted Instruction is defined as instructional activities that use a computer as the primary vehicle for teaching content or processes rather than one-to-one interaction with a student [75]. It is also known as Computer Aided Learning and Computer Assisted Learning.

Although this term encompasses almost all teaching modalities that use a computer, most published studies refer to 3D interactive packages or programs. Almost all studies reviewed by searching literature with the above terms, revealed that students consider the CAI as very interesting and dynamic study tools [76 - 80].

A study conducted in Mexico showed students who studied anatomy with the help of CAI had better performance compared to students who used the traditional method alone [76].

A Hungarian study noted that students with CAI taught anatomy course had better results compared to students who didn't use CAI [77]. A study conducted in India to test the perception of students to CAI found majority of students do not perceive CAI as a modality that could substitute dissection although CAI could complement it [78].

Students' Perception about Assessment in Anatomy

Assessment of Anatomical Science Competence

Assessment in medical education, in general, is fundamental for evaluating quality of teaching-learning outcomes, certifying competence, and ensuring safe medical practice [81, 82]. In anatomical study, the assessment should include all anatomical science sub-disciplines: macroscopic (gross), microscopic (cytology/histology), neuroscience (neuroanatomy), developmental (embryology), imaging (radiological), surface (topographical), living (non-cadaveric), non-living (cadaveric), practical (procedural) and clinical (applied) anatomy. The sub-disciplines can be assessed separately as individual discipline-based courses, collectively in a multidisciplinary systems-oriented course or in a problem-based fashion integrated with other basic science and clinical medicine subjects [83]. However, it is important to note that assessment of anatomical studies requires the evaluation of three scopes: theoretical, practical and clinical knowledge [84]. These scopes can be assessed by oral, practical or written methods [85].

Theoretical knowledge is usually assessed by written examinations in the form of multiple choice questions (MCQs) and short answer questions (SAQs). Practical knowledge is typically assessed by spot (*steeplechase) or* objective structured practical examinations (OSPE). Finally, clinical knowledge is generally assessed by objective structured clinical examinations (OSCE) [86].

Students' Perception about Formative Assessment

There are two broad types of assessments in anatomical science knowledge: formative and summative. These assessments, overall, provide improved self-efficacy, decrease extrinsic motivation and steer learning away from simply achieving test scores into an active and dynamic learning process [87]. Formative assessments take place regularly throughout the academic year. They mainly focus on qualitative feedback such as teaching-learning outcomes rather than merely grading scores [88]. Formative assessments are administered by instructors to monitor learning progress and provide ongoing constructive feedback about teaching-learning outcomes [89-91]. They are meant to identify students' strengths and weaknesses [92], guide future directions in learning and understanding [93], motivate students to celebrate their examination achievements [90, 92] and to promote self-motivation to academic development [94].

The sum of formative assessments may or may not be quantifiably graded and can constitute a portion of the students' final grade. When they are not quantifiably graded, students will have full freedom to experiment the subject in-depth, be more exploratory in their learning approaches, make associations across anatomical science disciplines and other sciences, and develop desirable higher cognitive skills without academic penalty [95]. Formative assessments without a grade can promote a deep learning approach [89, 96]. Formative assessments allow students to be actively engaged when they achieve the innate reward through advancing their knowledge, higher cognitive skills and overall course achievements [95]. Additionally, a bonus system can be incorporated into the formative assessment further strengthening the benefits of the assessment. [97].

Students can perceive formative assessments as stressful [98, 99] and a hindrance to smooth learning process [100, 101]. However, formative assessments being administered continuously on a regular basis tend to reverse this negative perspective over time [90]. Formative assessments prepare students to tacitly accept assessments as an inescapable component of curriculum [97] and an inevitable fact in medical education [90]. Therefore, eventually, students will have to adapt to the habit of assessment. Moreover, these formative assessments will impact positively on students improving their habitual studying mechanisms, gearing them towards key-learning matters, and providing them with regular feedback regarding their teaching-learning outcomes and how they can improve further [91].

Formative assessments can be administered through quizzes (daily/online/in-class), homework tasks, classroom participation, practical session involvement, laboratory worksheets, laboratory assignments, individual/group projects, mock OSCE/OSPE practice tests, mid-term/quarterly examinations [92], as well as peer-assisted evaluations, portfolios and self-assessments [93].

Students' Perception about Summative Assessment

Summative assessments are intended to confirm if the teaching-learning outcomes of a course have been met based on a professor's criteria. Summative assessments do not provide adequate feedback [102], and are perceived to be stressful as they examine the students' overall academic competence and accountability [93]. If not passed successfully, the applied grade may act as a hindrance to further advancement to higher levels of education [103]. To that end, students tend to adopt a strategic learning approach based on how they will be examined.

In anatomical learning, assessment of anatomical competence is largely measured using a hybrid of formative and summative grades in the theoretical, practical and clinical components [104-106]. Collective assessment of these three components provides an overall view about the student's anatomical competence [84]. However, how these respective domains should be assessed effectively and successfully is still debatable and demands further research [84, 104, 107, 108].

Studies have shown that regular formative assessments in the form of quizzes [90, 91, 104, 109, 110] assist in anatomical science learning [104]. Formative assessments, and particularly quizzes, have been shown to associate well with the written, practical and oral summative (end-of-year) assessments [90].

In a study by *Rowland* et al. (2011) [104], students identified practical assessments as the most recommended method for anatomy assessment, followed by written assessments and followed by oral assessments. Practical assessments such as OSPEs [111] promote a deep learning approach in which clinical anatomical knowledge and skills are integrated in a multidimensional fashion to enhance clinical decision-making [112].

Students Find Feedback after Assessment Useful

Assessment feedback by students should provide developmental growth as opposed to a judgment [113]. These feedback sessions are perceived as valuable and encouraging to create better teaching and learning [90]. While conducting frequent feedback sessions on assessment may be time and effort consuming [113], it can provide valuable information to improve anatomical learning [92].

The feedback sessions provide students an area to express personal views and constructive criticisms regarding the assessment in question [90]. The feedback sessions should be stimulating and support the spirit of clarification and learning from examination mistakes while not emphasizing the failed answers. Approaches that aim to maximize supportive sessions post-examination have been associated with better academic performance [114]. Student feedback sessions can be conducted in a variety of flexible ways such as one-on-one student interviews and structured group dialogues [115].

Alfaisal University Medical Students' Perception about Teaching and Learning Anatomy

Introduction to the Alfaisal University College of Medicine

The College of Medicine at Alfaisal University offers a seven-year MBBS undergraduate medical education program. The MBBS program is composed of three phases. Phase 1 is designed to introduce students to biomedical sciences.

Phase 2 marks the transition from basic biomedical sciences into system-based learning. Finally, phase 3 is the clinical clerkship phase where students apply their theoretical knowledge and practical skills into clinical practice.

Alfaisal University follows an innovative student-centered curriculum which is based on the "SPICES" Model, i.e., Student-centered, Problem-based, Integrated, Community-oriented, Elective and Systematic [116]. Partners Harvard Medical International (PHMI) currently collaborates with Alfaisal on the development of the curriculum, design of the facilities, organization of the administrative structure and development of the faculty.

At the College of Medicine — Alfaisal University, anatomical sciences are moving away from traditional subject-based courses to problem-based curricula. Furthermore, anatomy is also taught through the integrated resource center. This center is meant to address the aimed learning objectives in an integrated manner with dissected specimens, plastinated models, radiological images (X-ray, CT scan, MRI, Ultrasound), histopathological sections, audiovisual materials (surgeries and laparoscopies), surface anatomy (living human bodies), as well as clinical case studies with a set of questions and answers to self-test knowledge and foster understanding.

Anatomic study is assessed in a formative and summative manner at Alfasial University. Problem-based learning sessions contribute to the overall course grade. Examinations are implemented in the assessment of anatomical sciences which include MCQs, SAQs and OSPE, all of which are United States Medical Licensing Examination (USMLE)-formatted.

Methods

An online survey was structured and students' responses were collected and analyzed. The survey was structured using the Likert scale (from 1 strongly disagree to 5 strongly agree) to measure students' perception about anatomy. Students responses were categorized into three items which include agree (4+5), neutral (3) and disagree (1+2).

Results

Survey's Results

Among 530 undergraduate medical students, 33.20% (N=176) medical students, of which 51.70% (N=91) male and 48.30% (N=85) female students, responded to the survey. N values for the different levels are 48.30% (N=85) of first year medical students, 26.70% (N=47) of second year medical students, 11.36% (N=20) of third year medical students, and 13.63% (N=24) from fourth and fifth year medical students.

Majority of Alfaisal Medical Students Prefer Learning Anatomy via Dissection

Students were asked to select the most important method of learning anatomy from a list of various methods. They were also asked to rank group of means for learning anatomy in a descending order from the most effective to the least effective. (Figure 1)

Majority of Alfaisal's Medical Students Think That Anatomy Review Sessions Are Required More Often

Students were asked if they require anatomy review sessions more often. 77.27% (N=136) think that more frequent review session would be beneficial with the remainder, 22.73% (N=40) do not.

Majority of Students Think That Integrated Anatomy Teaching Is a More Effective Approach

Table 1 shows, through a Likert-Based questionnaire, that Alfaisal medical students agree that the integration of clinical concepts with other medical sciences assisted in understanding the concepts being presented.

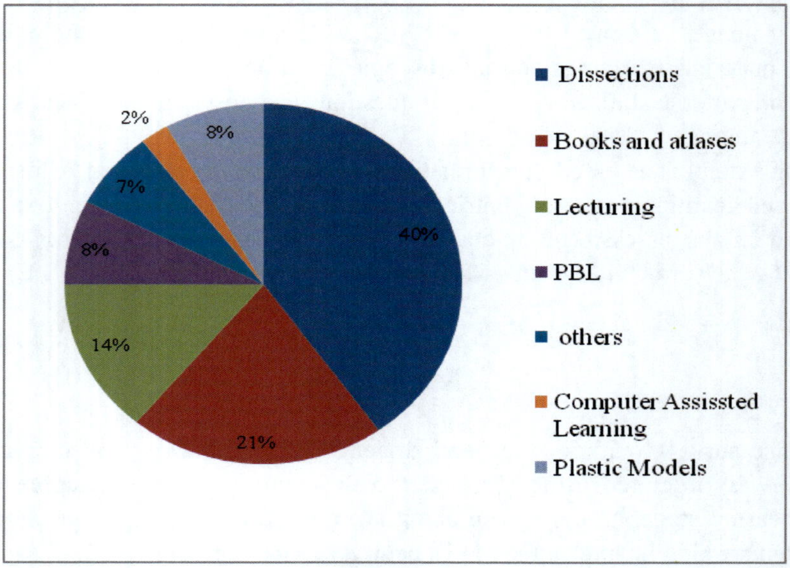

Figure 1: Percentage of students' choices of the most important learning method of anatomy.

Majority of Alfaisal's Medical Students Think That Their Anatomy Assessment is Structured in an Integrative Manner

Students were asked whether their assessment is based on testing their ability to integrate anatomy with other basic medical sciences. Results showed that 50.00% (N=88) agree that their anatomy assessment is structured in an integrative manner, while 23.29% (N=41) disagree, and 26.70% (N=47) are neutral. (Table 1)

Literature Review Results

Medical Students at Alfaisal University Expand Their Dissection Experience Beyond Anatomy

Undergraduate medical students at Alfaisal University were able to publish two case reports in the International Journal of Angiology. These two reports generated by students revealed two abnormal findings and pathologies. The first one, published in2011, was entitled "Persistent Bilateral Sciatic Artery: Report of a Rare Case". In this article the authors elaborate that the students were able to identify a persistent bilateral sciatic artery as unexpected abnormal finding during cadaveric dissection.

Table 1. Students' responses regarding learning anatomy in an integrative approach

Category and Evaluative items	Disagree N (%)	Neutral N (%)	Agree N (%)	Mean
I think the way we are taught anatomy is ineffective. Detail teaching is irrelevant to clinical practice	86 (48.86)	49 (27.84)	41 (23.29)	2.648
I understand anatomy better when it is integrated with physiology, histology, and embryology	30 (17.05)	33 (18.75)	113 (64.21)	3.761
Anatomy becomes more interesting and easier when put into clinical implication	10 (5.68)	24 (13.64)	142 (80.69)	4.267
I think by integrating different radiological modalities in teaching anatomy (X-ray, MRI, CT, ultrasound, etc) make anatomy teaching more effective	28 (15.91)	38 (21.59)	110 (62.50)	3.716
I think we should have more laboratory sessions	24 (13.64)	42 (33.86)	110 (62.50)	3.795
I think we should have more lecturing sessions	59 (33.52)	62 (35.23)	55 (31.25)	3.017
Our assessment focuses primarily on testing how I can integrate my Anatomy knowledge with other sciences	41 (23.29)	47 (26.70)	88 (50.00)	3.312

Sciatic artery is originally present during early development in-utero to supply lower limbs and then regresses once replaced by the inferior and superior gluteal arteries. This congenital anomaly stimulated the students to integrate their embryological knowledge with their gross anatomical knowledge. Furthermore, the finding shed light on noteworthy clinical significance because it is likely to be associated with aneurysm and causes obstacles during orthopedic hip surgeries and transplant surgeries. The author students concluded that this finding is of great importance to support the value of dissection with the ongoing debate on its efficacy [117].

The second article was published in 2012 and entitled as "How the Dissection Laboratory Facilitates Integration of Learning: Presence of Abdominal Aortic Aneurysm with a Large Intracardic Thrombus - A Rare Cadaver Finding". The students reported that this particular case was very interesting to them as the cadaver harbored more than one abnormality. The students were able to identify a big aortic aneurysm, an intracardic thrombus, absent left kidney, surgical aortic stent, and collaterals supply in thorax, as well as being able to assess the thrombus on a histopathological slide. This was an interesting case because the students were able to integrate their physiological knowledge of clotting cascade, pathology of aortic aneurysm, surgical tools and histology findings during their dissection. The students dealt with this case as a patient, elaborating many differentials in this case including cardiac tumor, protein C deficiency and deep vein thrombosis. They also sent a sample of thrombus for investigation and showed population of neurtophils. The author students once again concluded that dissection remains a significant modality in learning clinical anatomy [118]. These two case reports also contributed to improve students' research writing and thinking skills as due to this dissection-related finding, they were exposed to critical reasoning and writing early on in medical school.

Student-Centered Integrated Anatomy Sessions at Alfaisal University

This study was conducted to assess the efficacy of the integrated resources sessions (IRS) implemented at Alfaisal University. The students conducted sessions in the dry and wet laboratories concurrently, two hours, twice per week. Instructors helped groups of students to perform dissection of the ongoing theme of the curriculum and explain all related procedures and materials. These students then conducted 15-30 minutes session to explain anatomical landmarks and structures visible on cadavers to their peers using dissection and prosection.

The students then have sessions within the dry laboratory where they work in groups studying anatomy using histology slides, radiological modalities, clinical case scenarios, videos and plastinated specimens. Based on first-year medical students' responses to the structured questionnaire, the IRS seems to be a more attractive environment for first-year medical students to learn anatomy. The students also perceived the IRS improved their knowledge in a self directed environment. The majority of students reported that IRS helped them formulate their leaning objectives and served as a good resource for exam preparation. The majority, however, still believe that the fundamental role of lecturing and textbooks cannot be replaced by IRS [119].

PBL Performance Correlates with Assessment Content Acquisition: A Study in a Hybrid PBL Program at Alfaisal University

A group of medical education experts from Alfaisal University examined if there is a correlation between students' performance and assessment of content acquisition.

They ran analysis of 54 students who performed written examinations in semester three and their overall scores on PBL quizzes. It was found that there is a strong correlation between students' scores on written examinations and PBL performance using Pearson product-moment correlation coefficient test holding a value of $r=72$ ($p<0.0001$). The authors also described that there is significant correlation between the PBL performance and the different parts of the written examination which includes MCQ, SAQ and OSPE with r values equal to 0.59, 0.49 and 0.53, respectively [120].

Discussion and Conclusion

The primary objective of the conducted survey was to assess students' perception about learning and teaching anatomy at Alfaisal University College of Medicine. The questionnaire focused primarily on two important areas: determining the most useful modality for students to learn anatomy and students' perception about learning anatomy in an integrated fashion. Majority of students preferred learning via dissection which supported studies conducted in other institutes worldwide [121-123].

Currently, cadaver dissection serves the essential purpose of the ongoing reform in medical education, which favors teaching anatomy in an integrated fashion [124]. Dissection allows student to go beyond only learning anatomy which was clearly reflected by the two case reports published by Alfaisal University. The use of cadavers provides the chance to integrate anatomy knowledge with other disciplines and subjects such as: physiology, embryology, histology and clinical concepts [118, 119]. The students favor hands on experience, which is reflected by the great number of requests for more laboratory sessions. The fact that dissection promotes hands-on experience and acquisition of skills is also supported by other studies [124, 125]. Moreover, surgeons advocate dissection because it orients the students with anatomical variations in blood supply, for example. In addition, dissection allows students to have the sense of touch and familiarity with 3D visualization unlike other modalities [19].

Cadavers, however, do have the potential to be problematic with their high costs and potential hazards. [126]. In the United Kingdom, new medical schools are reinstating cadaver studies while older institutions fail to use them. Students exhibited inadequate understanding of anatomical arrangement of the human body in their clinical studies when only non-cadaver learning was done [127, 128].

The majority of students also favored learning anatomy through integration, expressing greater interest in incorporating anatomical structures with clinically oriented cases, other basic medical sciences and other modalities such as radiology. By incorporating the two, the students move away from simple memorization and towards an understanding of the different modalities as well as their integration with one another.

Modifying the students' learning approach might lead to better anatomy retention and understanding [30]. The arguments which arise against teaching anatomy in the modern fashion seem to be adopted by traditionalists who as well argue that students should not play a role in curriculum design and plan [11, 13-15]. With the expansion of medical knowledge, this seems to be not applicable. Students need to be developed in an educational environment where they should become able to identify their own learning needs and objectives [15]. However, evaluation of embraced new modalities for teaching anatomy should be investigated roughly before and during application [119].

References

[1] Klement BJ, Paulsen DF, Wineski LE. Anatomy as the backbone of an integrated first year medical curriculum: design and implementation. *Anat. Sci. Educ.* 2011 May-Jun;4(3):157-69.

[2] Cooke M, Irby DM, O'Brien BC. Educating Physicians: A Call for Reform of Medical School and Residency. 1st Ed. Jossey-Bass; San Francisco, CA: 2010. p. 320.

[3] Snelgrove H, Familiari G, Gallo P, Gaudio E, Lenzi A, Ziparo V, Frati L. The challenge of reform: 10 years of curricula change in Italian medical schools. *Med. Teach.* 2009;12:1047–1055.

[4] Kim KJ, Kee C. Reform of medical education in *Korea. Med. Teach.* 2010;32:113–117.

[5] Edgren G, Haffling AC, Jakobsson U, McAleer S, Danielsen N. Comparing the educational environment (as measured by DREEM) at two different stages of curriculum reform. *Med. Teach.*2010;32:e233–e238.

[6] Rao KH, Rao RH. Perspectives in medical education 5. Implementing a more integrated, interactive and interesting curriculum to improve Japanese medical education. *Keio. J. Med.*2007;56:75–84.

[7] Khalid BA. The current status of medical education in the Gulf Cooperation Council countries. *Ann. Saudi. Med.* 2008 Mar-Apr;28(2):83-8.

[8] Al-Gindan YM, Al-Sulaiman AA, Al-Faraidy A. Undergraduate curriculum reform in Saudi medical schools. Which direction to go? *Saudi. Med. J.* 2000 Apr;21(4):324-6.

[9] Shahabudin SH, Mahmud MN, Lim KE. Changing a conventional curriculum in an established medical school. *Med. Teach.* 1988;10(2):149-57.

[10] Clark, W. (1983). Student Involvement in Curriculum Development: A Different View. Retrieved 2012 fromhttp://www.eric.ed.gov/ERICWebPortal/search/detailmini.jsp?_nfpb=trueand_andERICExtSearch_SearchValue_0=ED241406andERICExtSearch_SearchType_0=noandaccno=ED241406

[11] Henzi D, Davis E, Jasinevicius R, Hendricson W. In the students' own words: what are the strengths and weaknesses of the dental school curriculum? *J. Dent Educ.* 2007 May;71(5).

[12] Tang TS, Hernandez EJ, Adams BS. "Learning by teaching": a peer-teaching model for diversity training in medical school. *Teach Learn Med.* 2004Winter;16(1):60-3.

[13] D'Eon MF, Harris C. If students are not customers, what are they? Acad Med. 2000 Dec;75(12):1173-7. Review. *Pub. Med. PMID*: 11112713.

[14] Helen M. Shields, Vinod E. Nambudiri, Daniel A. Leffler, Chitra Akileswaran, Edith R. Gurrola, Rachel Jimenez, Amy Saltzman, Peter A. Samuel, Kara Wong, Augustus A. White III, Janet P. Hafler, Jane N. Hayward, Stephen R. Pelletier, Richard P. O'Farrell, Paola G. Blanco, Steven M. Kappler, and Roxana Llerena Quinn. Using medical students to enhance curricular integration of cross-cultural content. *Kaohsiung. J. Med. Sci.* September 2009 vol 25 No 9. pg 493-502.

[15] Atkins KM, Roberts AE, Cochran N. How medical students can bring about curricular change. *Acad. Med.* 1998 Nov;73(11):1173-6.

[16] Yiou R, Good enough D. Applying problem-based learning to the teaching of anatomy: the example of Harvard Medical School. *Surg. Radiol. Anat.* 2006 May;28(2):189-94.

[17] Mellon AF, Mellon J. Logical debate on problem based learning. *BMJ.* 2006 Mar4;332(7540):550-1.

[18] Albanese MA, Mitchell S. Problem-based learning: A review of literature on its outcomes and implementation issues. *Acad. Med.* 1993;68:52–81.

[19] Turney BW. Anatomy in a modern medical curriculum. *Ann. R. Coll. Surg. Engl.* 2007 Mar;89(2):104-7.

[20] Elizondo-Omana, R., Guzman-Lopez, S., and Garcia-Rodriguez Mde, L. (2005). Dissection as a teaching tool: past, present, and future. Anatomical Record. Part B, *New Anatomist,* 285(1), 11-15.

[21] Monkhouse, W. S., and Farrell, T. B. (1999). Tomorrow's doctors: today's mistakes? *Clinical Anatomy,* 12(2), 131-134.

[22] Muller S. Physicians for the twenty-first century: Report of the project panel on the general professional education of the physician and college preparation for medicine. *J. Med. Educ.* 1984;59 (11 Pt 2):1–208.

[23] Ostbye T, Fernando ML, Robinson M, Weston WW. Introducing problem-based learning into a traditional undergraduate medical curriculum. *Natl. Med. J. India* 1996;9:231–6.

[24] Al-Haddad MK, Jayawickramarajah PT. Problem-based curriculum: Outcome evaluation. *Med. Teach* 1991;13:273–9.

[25] Levine HG, Forman PM. A study of retention of knowledge of neurosciences information. *J. Med. Educ.* 1973;48:867–9.

[26] Patel VL, Groen GJ, Norman GR. Effects of conventional and problem-based medical curricula on problem solving. *Acad. Med.* 1991 Jul;66(7):380-9.

[27] Vernon DT, Blake RL. Does problem-based learning work? A meta-analysis of evaluative research. *Acad. Med.* 1993 Jul;68(7):550-63.

[28] Entwistle, N. (1997). Contrasting perspectives on learning. *The Experience of Learning,* 2, 3-22.

[29] Biggs, J. (1987a). Student Approaches to Learning and Studying. Camberwell, Victoria: *Australian Council for Educational Research.*

[30] Smith CF, Mathias HS. Medical students' approaches to learning anatomy: students' experiences and relations to the learning environment. *Clin. Anat.* 2010 Jan;23(1): 106-14.

[31] Khaki AA, Tubbs RS, Zarrintan S, Khamnei HJ, Shoja MM, Sadeghi H, AhmadiM. The First Year Medical Students' Perception of and Satisfaction from Problem-based Learning Compared to Traditional Teaching in Gross Anatomy: Introducing Problem-based Anatomy into a Traditional Curriculum in Iran. *Int. J. Health Sci.* (Qassim). 2007 Jan;1(1):113-8.

[32] Kerby J, Shukur ZN, Shalhoub J. The relationships between learning outcomes and methods of teaching anatomy as perceived by medical students. *Clin. Anat.* 2011 May;24(4):489-97.

[33] Cornwall J. The diverse utility of wet prosections and plastinated specimens in teaching gross anatomy in New Zealand. *Anat. Sci Educ.* 2011 Sep-Oct;4(5):269-74.

[34] Jack A, Burbridge B. The utilisation of radiology for the teaching of anatomy in Canadian medical schools. *Can. Assoc. Radiol. J.* 2012 Aug;63(3):160-4.

[35] Azer SA. Can "YouTube" help students in learning surface anatomy? *Surg. Radiol. Anat.* 2012 Jul;34(5):465-8.

[36] Older J. Anatomy: a must for teaching the next generation. *Surgeon.* 2004 Apr;2(2): 79-90.

[37] Patel KM, Moxham BJ. The relationships between learning outcomes and methods of teaching anatomy as perceived by professional anatomists. *Clin. Anat.* 2008 Mar;21(2):182-9.

[38] Azer SA, Eizenberg N. Do we need dissection in an integrated problem-based learning medical course? Perceptions of first- and second-year students. *Surg. Radiol. Anat.* 2007 Mar;29(2):173-80.

[39] Winkelmann A. Anatomical dissection as a teaching method in medical school: a review of the evidence. *Med. Educ.* 2007 Jan;41(1):15-22.

[40] "Dissection" Medline Plus-Medical Dictionary Merriam-Webster, 2011. Web. 8 May 2011. <http:// www.merriam-webster.com/medlineplus/dissection>

[41] Bay BH, Ling EA. Teaching of anatomy in the new millennium. *Singapore Med. J.* 2007 Mar;48(3):182-3.

[42] Coulehan JL, Williams PC, Landis DD, Naser C. The first patient, reflections and stories about the anatomy cadaver. *Teach Learn Med.* 1995; 7:61-6.

[43] Samalia L, Stringer MD. A dissecting competition for medical students. *Anat. Sci. Educ.* 2012 Mar-Apr;5(2):109-13.

[44] Kang SH, Shin JS, Hwang YI. The use of specially designed tasks to enhance student interest in the cadaver dissection laboratory. *Anat Sci. Educ.* 2012 Mar-Apr;5(2):76-82.

[45] Bataineh ZM, Hijazi TA, Hijleh MF. Attitudes and reactions of Jordanian medical students to the dissecting room. *Surg Radiol. Anat.* 2006 Aug;28(4):416-21.

[46] Houwink AP, Kurup AN, Kollars JP, Kral Kollars CA, Carmichael SW, Pawlina W. Help of third-year medical students decreases first-year medical students' negative psychological reactions on the first day of gross anatomy dissection. *Clin. Anat.* 2004 May;17(4):328-33.

[47] Bernhardt V, Rothkötter HJ, Kasten E. Psychological stress in first year medical students in response to the dissection of a human corpse. *GMS Z Med Ausbild.* 2012;29(1).

[48] Plaisant O, Courtois R, Toussaint PJ, Mendelsohn GA, John OP, Delmas V, Moxham BJ. Medical students' attitudes toward the anatomy dissection room in relation to personality. *Anat Sci. Educ.* 2011 Nov-Dec;4(6):305-10.

[49] Granger NA, Calleson D. The impact of alternating dissection on student performance in a medical anatomy course: are dissection videos an effective substitute for actual dissection? *Clin. Anat.* 2007 Apr;20(3):315-21.

[50] Hofer RE, Nikolaus OB, Pawlina W. Using checklists in a gross anatomy laboratory improves learning outcomes and dissection quality. *Anat. Sci. Educ. 2011* Sep-Oct;4(5):249-55.

[51] Vasan NS, DeFouw DO, Holland BK. Modified use of team-based learning for effective delivery of medical gross anatomy and embryology. *Anat Sci. Educ.* 2008 Jan;1(1):3-9.

[52] Michaelsen, L.K. and Sweet, M. (2008). The essential elements of team-based learning. New Directions for Teaching and Learning, Wiley Periodicals Inc, V 2008, Issue 116, Pp 7-27.

[53] Burgess AW, Ramsey-Stewart G, May J, Mellis C. Team-based learning methods in teaching topographical anatomy by dissection. *ANZ J. Surg.* 2012 Jun;82(6):457-60.

[54] Vasan NS, DeFouw DO, Compton S. A survey of student perceptions of team-based learning in anatomy curriculum: favorable views unrelated to grades. *Anat. Sci. Educ.* 2009 Jul-Aug;2(4):150-5.

[55] Vasan NS, DeFouw DO, Compton S. Team-based learning in anatomy: an efficient, effective, and economical strategy. *Anat. Sci. Educ.* 2011 Nov-Dec;4(6):333-9.

[56] Lufler RS, Zumwalt AC, Romney CA, Hoagland TM. Incorporating radiology into medical gross anatomy: does the use of cadaver CT scans improve students' academic performance in anatomy? *Anat. Sci. Educ*. 2010 Mar-Apr;3(2):56-63.
[57] Nyhsen CM, Steinberg LJ, O'Connell JE. Undergraduate radiology teaching from the student's perspective. Insights Imaging. 2012 Dec 7. [Epub ahead of print] *Pub. Med. PMID*: 23225253.
[58] Ganske I, Su T, Loukas M, Shaffer K. Teaching methods in anatomy courses in North American medical schools the role of radiology. *Acad. Radiol*. 2006 Aug;13(8):1038-46.
[59] Dettmer S, Tschernig T, Galanski M, Pabst R, Rieck B. Teaching surgery, radiology and anatomy together: the mix enhances motivation and comprehension. *Surg. Radiol. Anat*. 2010 Oct;32(8):791-5.
[60] Rengier F, Doll S, von Tengg-Kobligk H, Kirsch J, Kauczor HU, GieselFL. Integrated teaching of anatomy and radiology using three-dimensional image post-processing. *Eur. Radiol*. 2009 Dec;19(12):2870-7.
[61] Swamy M, Searle RF. Anatomy teaching with portable ultrasound to medical students. *BMC Med. Educ*. 2012 Oct 22;12:99.
[62] Gunderman RB, Wilson PK. Viewpoint: exploring the human interior: the roles of cadaver dissection and radiologic imaging in teaching anatomy. *Acad. Med*. 2005 Aug;80(8):745-9.
[63] Haubert LM, Way D, DePhilip R, Tam M, Bishop J, Jones K, Moffatt-Bruce SD. Surgeons as medical school educators: an untapped resource. *Anat. Sci. Educ*. 2011 Jul-Aug;4(4):182-9.
[64] Seyfer AE, Welling D, Fox JP. The value of surgeons teaching anatomy to first-year medical students. *Bull Am. Coll. Surg*. 2007 Oct;92(10):8-14.
[65] Haubert LM, Jones K, Moffatt-Bruce SD. Surgical clinical correlates in anatomy: design and implementation of a first-year medical school program. *Anat. Sci. Educ*. 2009 Nov-Dec;2(6):265-72.
[66] Are C, Stoddard HA, Northam LC, Thompson JS, Todd GL. An experience in surgical anatomy to provide first-year medical students with an early exposure to general surgery: a pilot study. *J. Surg. Educ*. 2009 Jul-Aug;66(4):186-9.
[67] Are C, Stoddard HA, Thompson JS, Todd GL. The influence of surgical demonstrations during an anatomy course on the perceptions of first year medical students toward surgeons and a surgical career. *J. Surg. Educ*. 2010 Sep-Oct;67(5):320-4.
[68] Laparoscopic Surgery Recording as an Adjunct to Conventional Modalities of Teaching Gross Anatomy Pananghat A. Kumar, Mark Norrish, *Thomas Heming Sultan Qaboos Univ. Med*. J. 2011 November; 11(4): 497–502.
[69] Fitzpatrick CM, Kolesari GL, Brasel KJ. Teaching anatomy with surgeons' tools: use of the laparoscope in clinical anatomy. *Clin. Anat*. 2001 Sep;14(5):349-53.
[70] Glasgow SC, Tiemann D, Frisella MM, Conroy G, Klingensmith ME. Laparoscopy as an educational and recruiting tool. *Am. J. Surg*. 2006 Apr;191(4):542-4.
[71] Marker DR, Juluru K, Long C, Magid D. Strategic improvements for gross anatomy web-based teaching. *Anat Res Int*. 2012;2012:146262.
[72] Jaffar AA. YouTube: An emerging tool in anatomy education. *Anat. Sci. Educ*. 2012 May-Jun;5(3):158-64.

[73] Fruhstorfer BH, Palmer J, Brydges S, Abrahams PH. The use of plastinated prosections for teaching anatomy--the view of medical students on the value of this learning resource. *Clin. Anat.* 2011 Mar;24(2):246-52.

[74] Latorre RM, García-Sanz MP, Moreno M, Hernández F, Gil F, López O, Ayala MD, Ramírez G, Vázquez JM, Arencibia A, Henry RW. How useful is plastination in learning anatomy? *J. Vet. Med. Educ.* 2007 Spring;34(2):172-6.

[75] Computer-assisted instruction. (n.d.) Mosby's Medical Dictionary, 8th edition. (2009). Retrieved December 21 2012 from http://medical-dictionary. the free dictionary. com/computer-assisted+instruction.

[76] Elizondo-Omaña RE, Morales-Gómez JA, Guzmán SL, Hernández IL, Ibarra RP, Vilchez FC. Traditional teaching supported by computer-assisted learning for macroscopic anatomy. *Anat. Rec. B New Anat*, 2004 278, 18-22.

[77] Gary K, Samuel AC, Gréta k. Computer-assisted learning in anatomy at the International Medical School in Debrecen, Hungary: A preliminary report. *Anatomical Sciences Education,* 2012.

[78] Rehman FU, Khan SH, Yunus SM. Students, perception of computer assisted teaching and learning of anatomy-in a scenario where cadavers are lacking. *Biomedical Research*, 2012 23, 215-218.

[79] Tam MD, Hart AR, Williams S, Heylings D, Leinster S. Is learning anatomy facilitated by computer-aided learning? A review of the literature. *Medical Teacher.* 2009 31, 393-396.

[80] WALSH RJ, BOHN RC. Computer-assisted instructions: a role in teaching human gross anatomy. *Medical Education*, 1990 24, 499–506.

[81] Ahmed K; Ashra Wan H; Hanna GB; Darzi A; Athanasiou T. Assessment of specialists in cardiovascular practice. *Nat. Rev. Cardiol.* 2009;6:659–667.

[82] Larsen DP; Butler AC; Roediger HL. Test-enhanced learning in medical education. *Med. Educ*. 2008;42:959–966.

[83] Drake RL. Anatomy education in a changing medical curriculum. *Kaibog Zassi.* 1999;74:487-90.

[84] Schoeman S; Chandratilake M. The Anatomy Competence Score—A New Marker for Anatomical Ability. *Anat. Sci Educ.* 2012 Jan-Feb;5(1):33-40. doi: 10.1002/ase.263. Epub 2011 Nov 29.

[85] Lukic IK; Gluncic V; Katavic V; Petanjek Z; Jalsovec D; Marusic A. Weekly quizzes in extended-matching format as a means of monitoring students' progress in gross anatomy. *Ann. Anat.* 2001;183:575–579.

[86] Yaqinuddin A; Zafar M; Ikram MF; Ganguly P. What is an objective structured practical examination in anatomy? Anat Sci Educ. 2012 Aug 28. doi: 10.1002/ase.1305. [Epub ahead of print]

[87] Shepard, Lorrie A. (2005). Formative assessment: Caveat emptor. ETS Invitational Conference The Future of Assessment: *Shaping Teaching and Learning.* New York, USA; October 10–11, 2005.

[88] Huhta A. "*Diagnostic and Formative Assessment*". In Spolsky, *Bernard* and *Hult, Francis M. The Handbook of Educational Linguistics*. Oxford, UK: Blackwell; 2010; pages 469–482.

[89] Rushton A. Formative assessment: a key to deep learning? *Med Teach.* 2005;27:509–513.

[90] Poljicanin A; Caric A; Vilovic K; Kosta V; MarinovicGuic M; Aljinovic J; Grkovic I. Daily mini quizzes as means for improving student performance in anatomy course. *Croat Med. J.* 2009;50:55–60.

[91] Fenderson BA; Fishback J; Damjanov I. Weekly mini-examinations (quizzes) based on extended-matching questions as a means for monitoring medical student performance. *Croat Med. J.* 1996;37:283–7.

[92] Ganguly, PK. Teaching and learning of anatomy in the 21st century: direction and the strategies. *The Open Medical Education Journal*. 2010;3:5-10.

[93] Epstein RM. Assessment in medical education. *N. Engl. J. Med.* 2007 Jan 25;356(4):387-96.

[94] Ben-David MF. The role of assessment in expanding professional horizons. *Med. Teach*. 2000;22:472-7.

[95] Brown S; Knight P. Assessing Learners in Higher Education. London: Kogan Page; 1994.

[96] Tang, KCC. Perceptions of task demand, strategy attributions and student learning. Research and Development in Higher Education. 1992;15:474-481.

[97] McLean M. Introducing a reward system in assessment in histology: a comment on the learning strategies it might engender. *BMC Med. Educ.* 2001;1:7. doi: 10.1186/1472-6920-1-7.

[98] McLean M. Student perceptions of their first and second year experiences of medical studies at the University of Natal. MEd Thesis, University of Natal, Durban, South Africa 2000.

[99] Olmesdahl PJ. The establishment of student needs: an important internal factor affecting course outcome. *Med. Teach.* 1999;21:174–179. doi: 10.1080/01421599979824.

[100] Newble DI; Entwistle NJ. Learning analysis and approaches: Implications for medical education. *Med. Edu.* 1986;20:162–175.

[101] Knapper, C.K. Understanding student learning: Implications for instructional practice. In Wright WA, editor. Successful Faculty Development: Strategies to improve university teaching. Bolton, MA: Anker Publishing Company; 1995; pages 58-75.

[102] Schuwirth L; van der Vleuten C. Merging views on assessment. *Med. Educ.* 2004;38:1208-10.

[103] Sullivan W. *Work and integrity: the crisis and promise of professionalism in Ame*rica. 2nd Edition. San Francisco: Jossey-Bass; 2005.

[104] Rowland S; Ahmed K; Davies DC; Ashrafian H; Patel V; Darzi A; Paraskeva PA; Athanasiou T. Assessment of anatomical knowledge for clinical practice: perceptions of clinicians and students. *Surg. Radiol. Anat.* 2011 Apr;33(3):263-9. Epub 2010 Nov 27.

[105] Jones LS; Paulman LE; Thadani R; Terracio L. Medical student dissection of cadavers improves performance on practical exams but not on the NBME anatomy subject exam. *Med. Educ. Online.* 2001;6:2.

[106] Burns ER. "Anatomizing" reversed: Use of examination questions that foster use of higher order learning skills by students. *Anat. Sci. Educ. 2010;3:330–334.*

[107] Clough RW; Lehr RP. Testing knowledge of human gross anatomy in medical school: an applied contextual-learning theory method. *Clin. Anat.* 1996;9(4):263-8.

[108] Craig S; Tait N; Boers D; McAndrew D. Review of anatomy education in Australian and New Zealand medical schools. *ANZ J. Surg*. 2010 Apr;80(4):212-6.

[109] Percac S; McArdle PJ. Anatomy teaching: students' perceptions. *Surg. Radiol. Anat.* 1997;19(5):315-7.

[110] Nagar SK; Malukar O; Kubavat D; Prajapati V; Ganatra D; Rathwa A. Students' perception on anatomy teaching methodologies. *Natl. J. of Med. Res.* 2012;2(1): 111-112.

[111] Chakravarty M; Latif NA; Abu-Hijleh MF; Osman M; Dharap AS; Ganguly PK. Assessment of anatomy in a problem-based medical curriculum. *Clin. Anat.* 2005;18:131–136.

[112] Erkonen WE, Albanese MA, Smith WL, Pantazis NJ. Gross anatomy instruction with diagnostic images. *Invest. Radiol.* 1990 Mar;25(3):292-4.

[113] Brown S. Assessment for learning. Learning and Teaching in Higher Education, Issue 1, 2004-05.

[114] Kitchen E; King SH; Robison DF; Sudweeks RR; Bradshaw WS; Bell JD. Rethinking exams and letter grades: how much can teachers delegate to students? *CBE Life Sci. Educ.* 2006 Fall;5(3):270-80.

[115] Edgren G, Haffling AC, Jakobsson U, McAleer S, Danielsen N. Comparing the educational environment (as measured by DREEM) at two different stages of curriculum reform. *Med. Teach.*2010;32:e233–e238.

[116] Kattan K, Kattan W, Yakinuddin A. A Model for Program Evaluation in a Student Centered Undergraduate MBBS program at Alfaisal University, retrieved 2012 from http://heic.info/assets/templates/ heic2011/papers/15-Khaled_Kattan.pdf.

[117] Cowan MM, Hamweyah KM, Sabbagh MD, Swaid BA, Alkattan AK, Ganguly P. Persistent bilateral sciatic artery: Report of a rare case. *Int. J. Angiol.* 2010Spring;19(1):e43-4.

[118] Alyafi MM, AlAmodi AA, Juurlink BH, Ganguly P. How the dissection laboratory facilitates integration of learning: Presence of abdominal aortic aneurysm with a large intracardiac thrombus: a rare cadaver finding. *Int. J. Angiol. 2012*; 21:77-80.

[119] Cowan M, Arain NN, Assale TS, Assi AH, Albar RA, Ganguly PK. Student-centered integrated anatomy resource sessions at Alfaisal University. *Anat. Sci. Educ.* 2010 Sep-Oct;3(5):272-5.

[120] Yaqinuddin A, Kvietys P, Ganguly P, Ikram F, Yaeesh S, Kattan W. PBL performance correlates with content acquisition assessment: A study in a hybrid PBL program at Alfaisal University. *Med. Teach.* 2012;34(1):83.

[121] Weeks SE, Harris EE, Kinzey WG. Human gross anatomy: a crucial time to encourage respect and compassion in students. *Clin. Anat.* 1995;8(1):69-79.

[122] Mutyala S, Cahill DR. Catching up. *Clin. Anat.* 1996;9(1):53-6.

[123] Rajkumari, A.B., K. Das, G.T.N. Sangma and Y.I. Singh. Attitudes and views of first year medical students towards cadaver dissection in anatomy learning. *Calicut. Med. J.*, 2008; 6(4): e2.

[124] Rajkumari, A.B. and Y.I. Singh. Body donation and its relevance in anatomy learning - A review. *J. Anat. Soc. India,* 2007; 56(1): 1-6.

[125] Parker, L.M.. What's wrong with the dead body? Use of the human cadaver in medical education. *Med. J. Aust.*2002; 176(2): 74-76.

[126] Jones, D.G. Reassessing the importance of dissection: A critique and elaboration. *Clin. Anat.* 1997;10:123-127.

[127] McLachlan JC, Bligh J, Bradley P, Searle J. Teaching anatomy without cadavers. *Med. Educ.* 2004 Apr;38(4):418-24.
[128] Rizzolo LJ, Stewart WB. Should we continue teaching anatomy by dissection when...? *Anat. Rec. B. New Anat.* 2006 Nov;289(6):215-8.

INDEX

B

C

F

G

H

I

J

K

L

M

N

O

Q

R

S